# Quality
# Without
# Tears

**Other McGraw-Hill books by Philip B. Crosby**

*Quality Is Free* (1979)

*Running Things* (1986)

*The Eternally Successful Organization* (1988)

*Let's Talk Quality* (1989)

*Leading* (1990)

# Quality Without Tears

## The Art of Hassle-Free Management

**Philip B. Crosby**

**McGraw-Hill, Inc.**

New York   San Francisco   Washington, D.C.   Auckland   Bogotá
Caracas   Lisbon   London   Madrid   Mexico City
Milan   Montreal   New Delhi   San Juan   Singapore
Sydney   Tokyo   Toronto

**Library of Congress Cataloging in Publication Data**

Crosby, Philip B.
  Quality without tears.

  Includes index.
  1. Quality assurance.   I. Title.
TS156.6.C764   1984        658.5'62        83 26810
ISBN 0-07-014530-X

NOTE: A sixty-minute audio program to accompany this book is now available. Ask for it at your local bookstore or phone toll-free 1-800-2-MCGRAW.

First McGraw-Hill paperback edition, 1995.

27 28 29 30 DOC/DOC 0 9 8 7 6 5 4 3 2 1 0    (HC)
10 11 12 13 DOC/DOC 0 9 8 7 6 5 4 3 2      (PBK)


ISBN 0-07-014530-X  (HC)
ISBN 0-07-014511-3  (PBK)


*The sponsoring editor for this book was James H. Bessent, Jr., the editing supervisor was David E. Fogarty, and the production supervisor was Donald Schmidt. It was set in Primer by Achorn Graphic Services, Inc.*

*Printed and bound by R. R. Donnelley & Sons Company.*

To my hassle-free grandchildren, Charlie and Becky

# Contents

# Preface

When I wrote *Quality Without Tears* in 1984, I had been running Philip Crosby Associates, Inc. (PCA) for five years. We had conducted educational programs for thousands of executives and managers. We were expanding into Europe and Asia, and had developed video products to teach all the employees of those executives and managers. The growth of the company made being its CEO a full-time job. PCA was really not a consulting company, it was an education organization. We did some consulting, but mostly we taught people to do properly all the things they should have been doing anyway. My thought was that it is hard enough to understand an organization when you live there let alone coming in and out as a consultant.

*Quality Without Tears* came into being as the result of my desire to make it clear that the key to quality improvement was understanding concepts, not implanting some complex system. I wrote it on the run, at home and while traveling, mostly using a portable typewriter since personal computers had not become common at that time. I wanted to emphasize that causing quality was an ordinary part of the management job, not something that a special group was supposed to, or could, accomplish. I emphasized that all of this could be accomplished without beating up on the employees; a hassle-free organization was possible. To explain all this, I followed what was a slightly different approach for me. Grids and stories were used for the purpose of showing management that they are the problem. The question was how to do this so it wouldn't just be a cliché. It was necessary to make a real impact on people who did real work for a living.

In the past several years, as the book has been translated into several languages, I have talked with readers all around the world concerning whether the book had achieved this goal. They all seemed to relate to the profile of a quality troubled company, and the chapter on demotivation as showing a common problem worldwide. Those with large company experience related to the "Lightblue" corporation's communication problems. And everyone liked the broader explanation of the Four Absolutes of Quality Management. However, I discovered that very few people in Latin America, Africa and Asia had ever heard of Dickens's story, "The Christmas Carol." The upside of this was that they thought I had invented a great story even though I had made it clear that Dickens was responsible. I used the tale as a basis for showing the permanent effects of sloppy management standards today. We even made a movie of it starring Ephram Zimbalist, Jr. The video is available from Films, Inc., in Chicago.

The need for management's attention to quality became apparent to them with the publication of *Quality Is Free* in 1979. A great deal of progress has been made in the years since then and many companies have regained market share and their reputation for quality. However now, in 1995, I see things swinging back to the "good old days" when quality belonged to the quality department. That is like thinking that financial management belongs to the finance department. The trend is toward finding a "system" that will cause quality when installed, and applying for an award to show that it indeed exists. However, as before, there are no systems that guarantee anything in management. What a wonderful world we could have if it were only possible to have it that way. But the reality is that concepts and comprehension are what produce desirable results.

Many books begin their life with a great reception and then trail off into oblivion. *Quality Without Tears* began modestly and has built up "legs" over the years. This has come about because people found that they could actually use it as a pragmatic guide for causing quality. It explains concepts in everyday terms, and provides stories that give examples relating to real life. Actually, ghosts of Christmas probably do not represent real life, but their message is accurate.

Philip Ruppel and Jim Bessent of McGraw-Hill are responsible for this new edition, and I appreciate their work. Hope you find it useful, and enjoyable.

*—Philip Crosby*
*Winter Park, Florida*

# Preface to the Original Edition

In February 1979, there was an ice storm in Connecticut. This is a common occurrence, one that the residents never learn to like but do learn to handle. They go on somehow.

In the process of "going on," I was standing on the platform at the Port Chester, New York, train station. Although not as elegant or well thought of as the Greenwich station, it had the advantage of being closer to my home. In addition, an empty train formed up there at 7:42 every morning. After one stop at Rye, where it filled to capacity, it thundered on into Grand Central Station in the best tradition of early radio shows.

Through fourteen years of commuting I had learned that this route was the best of all available worlds. It was not fashionable, but it did provide a seat and a reliable schedule. Each day I would arrive at my office at 8:45, having walked up Park Avenue from the station. That day I knew the walk would be cold and perilous. Just passing the Union Carbide building (now sold to Manufacturers Hanover) was a stability test in itself.

On the train I began to wonder whether this was to be my fate forever. All around me were the familiar faces of the last 2,500 trips, each one growing a little more resigned each year.

There is nothing more exciting than working in Manhattan, and the commute from Greenwich is about as good as you can get. But there is a regularity about it that begins to get to you. If, for some reason, the newspapers don't show up and the commuters have to entertain each other—well, that can be a tragedy. Speaking out loud on the train is something that is done only by tourists or one-time riders.

My thoughts kept returning to the condo we owned in Vero Beach, Florida. Shirley was down there at this moment, but I had to be in New York for the monthly general management meeting. It would be several days before I could escape to warmth for the long weekend. Then it would be off to Brussels for a week of sessions concerning the inner world of ITT.

I always felt that the only thing worse than going to all those meetings was not being invited to attend them.

On arriving at the office, I received a telephone call from the editor of a popular business magazine. He said that they had just read my book *Quality Is Free* and thought it was terrific. They were going to review it in the magazine and wanted some quotes from me, plus a photograph, to put in an accompanying article. I arranged compliance.

Here was the opportunity. A widely read favorable review of the book would cause executives of other companies to call me for advice or assistance. I was already getting one such call a week as it was. This was the sign I had been waiting for.

If I could just obtain five or six clients, I thought, I could help them install quality improvement, write a book a year, give a few lectures—and move to Vero Beach for good.

It would mean leaving ITT, where I had been treated very well. It would mean giving up a pension two years before early retirement was possible. It would mean no health insurance, no disability insurance, no life insurance.

However, I figured that all those things would take care of themselves. The urge to get out in the world and see whether I could get American management to see the light was just too strong. Quality was fast becoming a lost item because management was so incredibly naive about it. The "workers" were being blamed for what the management was causing.

I figured that the ITT'ers had taken it all to heart and were getting good results—so why not the rest of the community? It was now or never.

So I went downstairs and quit.

It took me two months to convince everyone that I was really serious about leaving and another two to actually leave. In the meantime, we sold the Greenwich house (the only investment I ever made any money on) and bought one in Winter Park, near Orlando. Vero, we decided, was just too far from everything.

But we did retain the property there at John's Island just for that reason.

On July 1, 1979, my first day of unemployment since 1950, we were floating around the swimming pool in the backyard. We had formed PCA, Inc., and set up my den as "world headquarters."

Several companies had called and asked me to come speak to their management meetings on the subject of quality. It seems that every major company takes the top executives away for a strategy session each year. They usually have a guest speaker—more than likely the one with the current hot book—and play golf.

I established a policy of charging for these speeches, which some companies took exception to but others considered normal. One executive noted that he was giving me the opportunity to speak to 150 potential customers. I said I would go to the meeting for free if all of them would.

The first calls for something substantial came on that July 1. Tennant and IBM both contacted me to begin a relationship that continues to this date. They both wanted to learn what they could do about quality improvement. I wanted to help them but didn't really know then how to go about it. Between us however it all worked out.

It became very clear to me that the way not to cause improvement was to get involved in the technical problems of the client companies. The work had to begin with management concepts and then proceed to implementation of a formally run quality improvement process.

I visited several IBM facilities, talking with everyone who would talk. This is a remarkable organization, full of people who have their own opinions about everything but have a common thread of loyalty and dedication to the company and its policies. My concepts of quality were quite different from those used inside IBM. As a result, we had some lively discussions. Today we are almost exactly in agreement, and I haven't changed a bit.

Tennant is a Minneapolis-based company, 1/260 the size of IBM, but with the exact same problems concerning quality. These problems were based on the conventional wisdom activities that were described in *Quality Is Free*. The awareness that the management had of these differences was, of course, what encouraged them to want to change in the first place.

However, wanting to change and having the information to change are not the same. So after much concentrated thought and visits to a dozen other companies, I came to the conclusion that we would have to have formal seminars.

Rollins College in Winter Park was kind enough to volunteer the use of a classroom in return for a modest donation. We scheduled three classes, one each month, for the last quarter of 1979. Eighteen people per class was the limit, because I feel you can't learn much if the student and intructor can't get to know each other. We later expanded that to twenty-two per class when we obtained our own facilities.

The first class contained people from ten different companies. It was laid out for one week. My son, Philip, who had just graduated from Bentley, came to Winter Park as the comptroller. Those financial duties included carrying material up to the classroom every morning and back every afternoon.

Shirley was the other worker. Both of them were receiving salaries: I was still unemployed and running out of funds.

By the time we had finished the year, there were eight of us in the company and we had obtained space in downtown Winter Park to hold classes and set up offices.

I am not going to take this preface through a blow-by-blow account of the last three years. We went through a depression in late 1982 and almost went under. There is a rule somewhere that says as soon as you finish your expansion, everything drops off the screen. However, the Lord is the owner of the company, and we struggled through.

At this writing, we have ninety-three employees, of whom twenty-four are "professional," and we occupy 68,000 square feet of space. We have had over 9,000 students through the college courses and have bought thousands of lunches. During this time, we have dealt with hundreds of companies all over the world and listened to thousands of reasons why quality is hard to get.

We have seen hundreds of successes. They vary in degree in accordance with the amount of hard work applied to the concepts and techniques we have learned. Nobody didn't improve.

We have learned how to consult, and how to guide the quality improvement teams so they don't fall into the bog along the way. We have seen some wonderful things happen. I believe that the country is beginning to become adult about quality. Some of that has to be due to our work.

But what makes me happiest is to look around and see PCA as a community of people. A company that came into being without the slightest help from the government. One that promises full employment to a lot of people forever if they only continue to produce. It lets you know that America still works.

This book is not a memoir. Someday I will write one of those and call it *Pin High in the Sand*. But this isn't it.

The purpose of this book is to help the reader understand more about quality improvement and the concepts that make it happen. It is an incredibly complex subject because it involves every single management action that is taken in an organization. Every decision, even one made in casual conversation, comes home to roost.

What originally began as "quality" has now expanded with experience into an understanding that "hassle elimination" and "quality improvement" are the same. So I have taken the basic elements of improvement—determination, education, implementation—and broken them down to their essence. Then there are case histories and essays to assist in understanding the impact management has on the company's results when it hassles the employee, knowingly or unknowingly.

This is by no means a cookbook. It is for comprehension and communication. It will help the reader understand what the battle is all about and what weapons are required for overcoming. Our experiences of the past four years provide an indication of the effort that must be put forth to cause the cultural revolution that is required.

I would like to thank Peggy Davis, my assistant, Betti Stalions, her secretary, and Joyce Kinney in word processing who patiently pulled the book together from my out-of-order inputs.

—*Philip Crosby*
*Winter Park, Florida*

# 1
# The Profile of a Problem Organization

When a physician sees a patient with red spots, a fever, and a brother who has the measles, it is not necessary to be Louis Pasteur to make a diagnosis. Identifying companies with big quality troubles has somewhat the same degree of difficulty.

Dissatisfaction with the final product or service of an organization is called trouble with quality. However, it is only a symptom of what is happening inside the firm.

There are several characteristics that troubled organizations have in common. Before launching into a discussion of the causes and cures of nonquality, we should examine some of the symptoms of the patient.

1. *The outgoing product or service normally contains deviations from the published, announced, or agreed-upon requirements.* Product companies have waivers, "off-specs," material review decisions, and so forth incorporated in outgoing material. This means that every unit is different. The "patients" see nothing wrong in this because they have carefully documented each nonconformance and ensured that it does not interfere with the form, fit, or function of the product. They fail to recognize that not only do they lose control of the outgoing product, but all that fooling around costs much more than it would to make everything as agreed.

   Service companies do not usually document their nonconformances as formally as product companies. However, bank statements, credit card bills, insurance policies, hotel reservations, and similar outputs routinely contain information that is not in accordance with procedural requirements. One insurance company I know of still misspells clients' names 24 percent of the time. The employees giggle about that.

The ease of living in a situation where nonconformance is the norm produces a consistent flow of problems. This consistency alone convinces everyone that "this is the way life is." Thus the situation feeds on itself. If things are always going to be that way, then it is necessary to take some other steps to ensure customer satisfaction. This leads us to the next symptom.

2. *The company has an extensive field service or dealer network skilled in rework and resourceful corrective action to keep the customers satisfied.* Product companies have "customer engineers" (CEs) who repair the copiers, computers, furniture, and other products that come to the customer directly. Many times the CE does the installation, which provides an opportunity to finish the product assembly in the customer's office without the customer's knowing what is going on. Little plastic bags with wires and notes in them are a tip-off. Customers love the CE, hate the company.

Companies with extensive dealer networks, such as automotive organizations, conduct "get ready" routinely. This means that the dealer finishes the product. The factory does not deliver a car (or similar product) that can be driven away from the end of the line and used. If you want to pick up a car at the plant, it must go through a different operation after it is built. Product recalls occur now and then to let the company finish something else. Most of these defects were known about beforehand—there have been very few surprises in product recalls.

Field service and dealer organizations consider themselves the vital connection between the company and the customer. When you realize that the customer could not use the product without their services, it is not hard to understand why they feel so important. In many companies the field service operation represents a large part of revenues because of its service contracts. Work done under the warranty is not profitable, however.

Service companies have their own way of providing field service. Credit card operations give a name and telephone number so that we will have someone to talk to when things go wrong;

banks provide a "personal banker" who runs interference and translates for you; insurance salespeople spend most of their time trying to make the "home office" get the customer's data right.

Hotels install "hot lines" so that when the staff fails, a guest can call an assistant manager who overrides the system and produces extra towels or whatever. Airlines dress employees in different colors so that the flyer can deal with the representative who handles specific problems.

All these actions represent a custom of patching that began a long time ago and is now deeply entrenched in the self-fulfilling prophecy that "this is the way life is." Individuals spend a lifetime with a company and retire content, never having done anything but rework.

When the service is expected to be incomplete and the product is assumed to always require some adjustment, a situation emerges in which the employees create their own performance standards. That leads us to the next symptom.

3. *Management does not provide a clear performance standard or definition of quality, so the employees each develop their own.* In product companies "quality standards" tend to be based on what the process itself actually produces. When the field finds 4 percent bad, the standard becomes "outgoing quality level of 4 percent defective." That sounds very precise and scientific. However, all it means is that the operation has settled on a level of incompetence.

"Schedule first, cost second, quality third" becomes the tradition, once employees see what happens to those who miss schedule or cost.

"Yield" is another expression used in processes. When the basic assumption is made that the process can never be operated without error, then the next step is a matter of putting an agreed-upon number on that error. If the yield is planned at 85 percent, that means a commitment to 15 percent error. People who are into "yield management" will tell you that isn't true, but it is.

When administrative companies establish off-line systems so customers or senior executives can bypass the bottlenecks, they make a clear statement. That statement is: "We don't really expect to meet our requirements, so just cope the best you can."

Thus employees are rewarded for resourcefulness. House newspapers proudly note cases in which employees knocked themselves out to get something done for a customer "in the best tradition of service." What is overlooked is that the knocking out would not have been necessary if the job had been done right the first time.

| Characteristic | That's us all the way | Some is true | We're not like that |
|---|---|---|---|
| 1  Our services and/or products normally contain waivers, deviations, and other indications of their not conforming to requirements. | | | |
| 2.  We have a "fix it" -oriented field service and/or dealer organization. | | | |
| 3.  Our employees do not know what management wants from them concerning quality. | | | |
| 4.  Management does not know what the price of nonconformance really is. | | | |
| 5.  Management believes that quality is a problem caused by something other than management action. | | | |
| | 5 Points | 3 Points | 1 Point |

Point count condition

| | | |
|---|---|---|
| 21 − 25 | Critical: | Needs intensive care immediately. |
| 16 − 20 | Guarded: | Needs life support system hookup. |
| 11 − 15 | Resting: | Needs medication and attention. |
| 6 − 10 | Healing: | Needs regular checkup. |
| 5 | Whole: | Needs counseling. |

The profile of a quality-troubled company.

The amazing part of all this fixing and reacting is that management doesn't realize what it has wrought in terms of expense. That brings up symptom number four.

4. *Management does not know the price of nonconformance.* Product companies, as we will see further along in the book, spend 20 percent or more of their sales dollars doing things wrong and doing them over. Service companies spend 35 percent or more of their operating costs doing things wrong and doing them over.

These expenses are very real and very high. A prevention-oriented quality management system can replace all that cost with the modest expense of an education and monitoring process.

If it is all so clear and obvious, how come management puts up with it? That brings up the fifth symptom, which is really the most important.

5. *Management denies that it is the cause of the problem.* The denial, for all types of businesses, is based on the random improvements that occur when any specific problem is attacked. Push in the balloon in one place, however, and the air goes to another area.

Most managers send everyone else to school, set up "programs" for the lowest levels of the organization, and make speeches with impressive-sounding words. It is not until all the problems are pulled together, particularly the financial ones, that the seriousness of the situation is exposed.

There is a parallel here with drug abuse, where the primary symptom also is denial. "I can handle it," is what victims all say. Usually they discover they can't handle it only when their life falls apart. With companies it occurs when the market share shrinks and the profits disappear. The main obstacle to improvement is the stubbornness of management.

# 2
# The Quality Vaccine

An organization can be vaccinated against nonconformance. It can be provided with antibodies that will prevent hassle. Some of these antibodies are managerial actions; some are procedural common sense.

For instance, the practice of qualification-testing newly developed products is something that every right-minded company routinely accomplishes. Service companies prove processes and practices before causing their implementation.

Just doing "provings" does not cause the proper actions to result, of course. Someone has to see to it that corrective action is taken, then follow up to ensure that all the safeguards are exercised. It is by not doing what they already know they should do that companies get into trouble over quality.

Consider the financial controls and guidelines that exist in well-conducted businesses. There are definite rules about how much money can be borrowed, how much should be spent by this or that function, and even about where corporate funds should or should not be invested. Perhaps more books are written about financial management than about any other subject (except, of course, dieting).

Yet each day *The Wall Street Journal* reports several financial disasters, real or in the making. Each week or month business magazines go into greater detail as to how the failures (or recoveries) occurred. In most cases, some basic rule of business management was violated or altered.

It is usually fairly clear where the responsibility originated. The managers were not stupid or evil people, however; they made a judgment and it didn't pan out. They must pay the penalty. Sometimes the wrong person is given the cup of hemlock, but that is financial life.

The typical management decision that causes quality problems results from a "hunch." It is a short-range solution to some problem of schedule or cost, or it happened because someone just didn't want to believe the results of evaluation.

The antibodies for these and similar bacteria are not found in detailed procedures and controls. Controls help, but it is very difficult to make motivated management people read the procedures, let alone follow them. Having a large book of policies and practices never saved any company from disaster.

Antibodies must be built into the management style that operates the company. The knowledge that they exist has to extend throughout the organization. This extension serves to help those who make decisions (and every employee makes decisions) to keep from causing problems of quality.

All nonconformances are caused. They are not genetically produced, they do not float in through the office window, they do not hide in the wrapping paper. They are caused. Anything that is caused can be prevented. The organization that wishes to avoid internal hassle, eliminate nonconformance, save itself a bundle of money, and keep its customers happy must be vaccinated.

To prepare the vaccine, you need to combine certain key ingredients (see pages 8–9). To administer it continually to the corporate body requires a strategy that contains three distinct management actions:

- Determination
- Education
- Implementation

Determination evolves when the members of a management team decide that they have had enough and are not going to take it any more. They recognize that their action is the only tool that will change the profile of the organization.

Education is the process of helping all employees have a common language of quality, understand their individual roles in the

(*Text continues on page 10.*)

# THE CROSBY VACCINATION SERUM INGREDIENTS

## Integrity

A. The chief executive officer is dedicated to having the customer receive what was promised, believes that the company will prosper only when all employees feel the same way, and is determined that neither customers nor employees will be hassled.

B. The chief operating officer believes that management performance is a complete function requiring that quality be "first among equals"—schedule and cost.

C. The senior executives, who report to those in A and B, take requirements so seriously that they can't stand deviations.

D. The managers, who work for the senior executives, know that the future rests with their ability to get things done through people—right the first time.

E. The professional employees know that the accuracy and completeness of their work determines the effectiveness of the entire work force.

F. The employees as a whole recognize that their individual commitment to the integrity of requirements is what makes the company sound.

## Systems

A. The quality management function is dedicated to measuring conformance to requirements and reporting any differences accurately.

B. The quality education system (QES) ensures that all employees of the company have a common language of quality and understand their personal roles in causing quality to be routine.

C. The financial method of measuring nonconformance and conformance costs is used to evaluate processes.

D. The use of the company's services or products by customers is measured and reported in a manner that causes corrective action to occur.

*E.* The companywide emphasis on defect prevention serves as a base for continual review and planning that utilizes current and past experience to keep the past from repeating itself.

## Communications

*A.* Information about the progress of quality improvement and achievement actions is continually supplied to all employees.

*B.* Recognition programs applicable to all levels of responsibility are a part of normal operations.

*C.* Each person in the company can, with very little effort, identify error, waste, opportunity, or any other concern to top management quickly—and receive an immediate answer.

*D.* Each management status meeting begins with a factual and financial review of quality.

## Operations

*A.* Suppliers are educated and supported in order to ensure that they will deliver services and products that are dependable and on time.

*B.* Procedures, products, and systems are qualified and proven prior to implementation and then continually examined and officially modified when the opportunity for improvement is seen.

*C.* Training is a routine activity for all tasks and is particularly integrated into new processes and procedures.

## Policies

*A.* The policies on quality are clear and unambiguous.

*B.* The quality function reports on the same level as those functions that are being measured and has complete freedom of activity.

*C.* Advertising and all external communications must be completely in compliance with the requirements that the products and services must meet.

quality improvement process, and have the special knowledge available to handle antibody creation.

Implementation is guiding the flow of improvement along the "yellow brick road." This process never becomes complete because the body continually changes. But each step along the way contributes to the health of that body.

Each of these subjects is discussed in detail in its own separate chapter.

Most companies concentrate on implementation before they do anything about the other two. This comes about because senior managers usually do not know much about quality except that "if you spend too much on it, you'll lose money, and if you don't spend enough, you'll lose money."

So when problems arise, it is entirely normal to reach out for the current fad. At this writing *quality circles* and *statistical quality control* meet that definition. Some companies have been using these two techniques, all by themselves, for several years and are discovering that (1) they are very hard to keep going and (2) very little actually improves.

There is nothing wrong with QC and SQC. They are excellent tools in the battle for quality improvement, but they are only tools. They are not management tools, yet they do not work unless management becomes completely involved. I meet few managers who understand even how these techniques work, let alone how to properly implement them.

To dehassle forever, it is necessary to change the company's culture, to eliminate the causes that produce nonconforming products and service. Let me provide a personal example.

Several years ago I had a heart attack in the middle of the night. My wife drove me to the hospital, where I was placed in intensive care and stabilized. After a few anxious hours it became apparent, although not to me, that it was going to turn out all right.

The next afternoon my family doctor came to see me. After inquiring about my condition, he said that he wanted to talk to me and he wanted my attention.

"Arthur," I said, "there will never come a time when you will have my attention any more fully than this."

He didn't seem impressed.

"Every year you come to me to get a physical examination," he began. "And every year I tell you that you have to lose 20 pounds and quit smoking.

"Every year you tell me how you are going to do something about it, and when you return the following year, it is apparent that you have done nothing.

"Now what I want to tell you is this: if you don't quit smoking and lose 20 pounds, you are going to die."

"Oh," I said.

This really impressed me. Arthur wasn't talking about statistics; he was pointing his finger straight at me. He hadn't commented about the probability that a person of my age, situation, and so forth would have this problem; he was talking about ME.

So I took it to heart. When my hospital stay was over, I continued not smoking and to this day have had no urge to return to that habit. Twenty years of something I once enjoyed went right out of my head.

Arthur wanted me to go from 195 pounds to 175 pounds. As it turned out, that was no problem. All I did was pick out one of the diet books that every bestseller list contains and apply it. Shirley guided me in my complete obedience to the techniques, and within three weeks I weighed 175 pounds.

I could only wonder why I had fought this for so long. It really felt great to have that excess flab gone. My size 43 suits were converted to size 41 by the tailor, and those that couldn't make it were replaced. I went back to work running around the world, trying to make ITT safe for quality.

Six months later I weighed 195 pounds again. Except now my suits didn't fit. It was very embarrassing.

After some thought it became apparent to me that the principles of quality management I spent my time preaching about were somehow missing from my actions in this situation. A short-range solution had been applied, and short-range results had been achieved.

It was not some conspiracy that was making me fat. No one was sneaking into my room when I was asleep and pouring food in my ear. It was not my wife or my boss or my staff. It was me. I ate too much.

In an effort to turn weight loss into a manageable project, I selected a product: a 175-pound Phil Crosby. All that was needed was to determine the requirements that would produce such a product and develop the management style that would maintain it.

So I kept track of everything that entered my mouth for two weeks. If you think you don't eat much, write it all down for fourteen days. This produces a very long list that would convert into an enormous pile of material.

My accounting revealed that control was lacking. There was entirely too much consumption going on, much of it thoughtless. It was obvious that there were no clear requirements involved in the eating process.

I defined the following requirements:

1. Have three meals a day only, no snacks.
2. Have one helping per meal.
3. Put the fork down between bites. (A psychologist friend told me this lets you know when you are full.)
4. Walk a couple of miles a day.

One thing we learned was that there was just too much food around. The kids had left to set up their own homes and we were still cooking for four. We were putting food in bowls and taking it to the table. When the bowls were empty the meal was over. This was changed so that the plates were served in the kitchen and the remaining food was disposed of. We also found that our dinner plates were very large, and we obtained smaller ones.

Another item that had to be dealt with was the business of eating out. We liked to walk over to Winter Park for dinner, and there were a lot of banquets and dinners involved in the course of work. So a new requirement came into being.

5. When dining out, only eat half of what was served.

This combined effort produced a weight reduction of 7 pounds over the following three weeks. Then the rate of loss flattened out and no more reduction came. This showed that the requirements so far defined produced a lifestyle that caused a 188-pound Phil Crosby. We had to do better than that.

Another review of the intake list revealed that the next big item was desserts. It was obvious that they had to be eliminated, but I really didn't want to give up ice cream altogether. By experimentation it was determined that I could have one hot fudge sundae every two weeks. All other desserts, including the chocolates in Brussels, were eliminated. That really hurt.

These new requirements dropped the product down to 182 pounds. Then the loss rate again flattened out and would not move. It was very tempting to stop there; the suits fit well and I felt fine. So I called Arthur and told him how things were going. He was delighted with my progress but would not retract his prognosis. Additional management-style weight management was going to be required.

The only item left on the food list was alcohol. I didn't drink beer or whiskey but I did drink wine. A lot of wine, as a matter of fact. In traveling a third of the time, I had become adjusted to the customs of the international world. One of the most common of these customs is the enjoyment of wine with meals. Wine contains alcohol, which is sugar and makes you fat. So the wine intake had to be reduced.

I decided to have only two glasses (small) of wine with meals and none otherwise. This worked out fairly well except that my brain apparently contains a relay that is shorted out in the middle of the second glass. The effect of that short is to dim the significance of the two-glass requirement.

All this became such a hassle that I decided to just eliminate alcohol completely. Without alcohol I lost the remaining 7 pounds, slept better, and found that time changes didn't bother me nearly so much. That was the good news. The bad news was that there was no one to talk to after 8 p.m. when I was traveling.

This discipline has had its effect. There is no need for a diet now, only for conformance to the requirements.

Getting weighed ten times a day in different ways would not reduce the weight a bit. Chopping off a leg would reduce the weight, but it would not produce a whole Phil Crosby. The only way the food gets into the body is by the actions of the body. Prevention is the act of not putting the food into the body.

When my daughter got married, I was handed a glass of wine during the reception. "What kind of world is it," I thought, "if a man can't have a sip of wine on his only daughter's wedding day?"

But I put it down and toasted the couple with water. My decision to conform to the zero-defect requirements (ZD) had been made some years before. If I broke that decision then, it would soon be my wedding anniversary, then it would be my birthday, and after that just the fact that it was Friday would be enough reason. My decision would have to be made all over again. Reaching the desired weight was a matter of culture change. Keeping it there is a matter of management style.

The determination, education, and implementation chapters will show what must be done to eliminate hassle and install defect prevention.

# 3
# Demotivation

Managers worry about motivation. Getting employees "turned on" has become a major industry. Speakers on the subject offer tapes, films, counseling, and many other types of support. For the most part the material is useful and the individuals are sincere. People go into an upbeat phase as a result—at least for a while.

However, we must ask ourselves, "Why do we need a special program to motivate our people? Didn't we hire motivated employees?"

If you think about it, they were well motivated when they came to work. That first day, when they reported in, there were nothing but smiles. A little nervousness, perhaps, but that is a natural and a desirable reaction. Everything was positive.

The individuals were determined to make a good impression, to learn the job well, to perform to the top of their ability. Perhaps they were ambitious and saw this first slot as the path to the CEO's job. Perhaps they were not all that determined but looked at it as a new start and a great opportunity.

At any rate, it is certain that they looked very positively at the company that was taking them on. Their attitude was good, they were paying attention, they took the whole situation very seriously.

However, a few months or years later things are different. The employee is not all that thrilled with the company and the job. The personal performance standard that was set at full throttle in the beginning now slumbers along. All the clichés have become part of the conversation:

"Don't make waves."

"No one knows what is going on around here."

"Why are they doing a dumb thing like that?"

"It's who you know, not what you know."

"They don't care about quality."

"There is no way to get ahead around here."

"I had this good idea but nothing happened."

As a general malaise sets in, someone in management decides that the employees need some better communication and support. This is usually a genuine approach with the theme of "making this a great place to work." A search is conducted for programs to use so that the employees can be trained to communicate.

The result of this search is "quality of work life" or "quality circles" or productivity improvement or something similar. There is nothing wrong with these programs except that they are all aimed at the bottom of the organization.

Reflect back to your childhood days when there was a school playground where a lot of fun games could be played and pleasant times had. Then introduce a couple of oversized bullies into the situation. These characters make it difficult for the rest of the people to play their games or even enjoy themselves. They work to no known rules and terrorize on a random basis. They interrupt the game, take the ball, and do whatever it is that turns them on.

Suppose, in order to overcome this problem, the school executives decide to install a motivation program. However, they send the victims to school, not the bullies. That is what happens in business.

Employees are turned off to the company through the normal operating practices of the organization. The thoughtless, irritating, unconcerned way they are dealt with is what does it. They feel they are pawns in the hands of uncaring functional operations.

Unfortunately, it only takes a few months of living at the top of the organization for a person to forget all this. So very little gets corrected. That is why revolutions usually fail to accomplish their objectives. (The French Revolution in 1789 produced Napoleon.)

What sort of indignities are we talking about? We don't need to reach back to Charles Laughton mismanaging the *Bounty* for our example. Certainly there hasn't been a case of "keelhauling" in

American industry for decades. The indignities that cause the problems are much more subtle than that. Let's consider a few typical turn-offs.

The performance review, no matter how well the format is designed, is a one-way street. Someone the individual didn't select gets to perform a very personal internal examination. There are no certificates hanging on the wall stating the qualifications of the reviewer. Yet the effect on the individual's present and future is as real as if everyone knew what he or she were doing.

Now this might lead one to think that the reviews are therefore very negative and have the effect of doing bad things to people. Once in a while that happens, but the overwhelming bundle of performance reviews are kind and gentle. They provide raises, recommend promotions, and in general exude goodwill.

If there is doubt about this, just examine the reviews of every single secretary in any corporate headquarters. There are none who are less than above average, and all are recommended heartily. I guarantee that.

The result of all this is to make the reviews counterproductive. They not only don't weed out the bad; they don't bring the good to the surface. And that is what drives people nuts.

Dishonest evaluations show people that the company has no integrity, doesn't trust a system it forces on them, and doesn't really care about finding talent if it exists. They feel ignored and abused. The reviews, which are supposed to give management information about the employees, do the reverse. The employees quickly realize that management has no way of knowing who is the fairest of them all, except through luck and instinct.

This adds discouragement to the list. The truly valuable begin to look around outside soon after their first evaluation.

Expense accounts are another prime opportunity to convince employees that their concern is of no concern. Most companies have no problem reimbursing their traveling employees for rooms, meals, and airplane fares. All these—which probably make up 97 percent of the total amount—are costs the traveler can't do much about. Meals vary somewhat, but the hotel prices are usually accepted.

The hitch arrives in writing off the items that someone can make a rule about, such as:

- Laundry may be sent out only on trips exceeding three days.
- Long-distance telephone calls home may be charged only every two days.
- When meals are served on an airplane, the traveler may not charge a meal upon arrival.
- Receipts are required for all automobile tolls, including parkways.
- No travel money may be drawn in advance.

The result of these edicts is an eternal battle between the accounting department and the travelers. Accounting gets to determine whether the traveler has complied or not. If the accounting people ever lose an argument, they just redesign the form or restate the rules.

The traveler must submit an expense account up into the organization, through a couple of bosses. The secretary has to make it out because the traveler cannot understand it. The difficulty of the form is a key to the success of the mission. The mission is to make traveling difficult.

The staunch ally of the expense account is the travel request. No one can just pick up the telephone and order a ticket, for obvious reasons. Tickets are real money, and real money has to be dispensed in an orderly fashion or a company can get into serious financial problems.

Many companies find a way to make this orderly process degrading. One sure-fire technique is to have different levels of people travel in different classes of service. This lets everyone know who is important and who isn't. The salary grade that marks the beginning of first-class travel privileges becomes the only significant one in the company. Continuing battles are fought to have subordinates raised to that level.

A particular humiliation occurs when one who must fly tourist accompanies one who rates first class. The lesser person can be upgraded if a humble enough letter is written to the director of administration and concurrence achieved.

Meetings also orchestrate themselves to cause demotivation. There seem to be certain aspects of meeting behavior that apply to all I have ever participated in. There is a constant number of speakers in a meeting, and that number doesn't change with the size.

Fewer than a half dozen, and usually three or fewer, will dominate the meeting. One is trying to find out what is going on, and the others are working either for or against that.

One older executive, who just joined a company I was working with, commented to me: "Staff meetings are the same everywhere. The boss talks about what he wants to talk about as long as he wants to talk about it and then the meeting is over."

There is absolutely nothing more demotivating or demeaning to a budding executive than to have to go to meetings where the assigned role is to be a faithful listener. We all know that the higher one is in the organization, the less actual information one possesses about a subject. That is not a cynical view at all, just practical. Senior executives have to cover a broad area of activities. It is not possible to know them all.

However, this doesn't stop them from having opinions and desires. They know what they would like to have happen and push for this, perhaps unconsciously, by leading the conversation in that direction. Meanwhile, those who really know must squirm uncomfortably in their seats. When told this, the senior person will smile patiently and point out that all the attendees have the opportunity to speak up whenever they want. Why, there is even a time at the end of the meeting when they go around the table to make certain nothing has been overlooked.

Yet, if a time usage ratio is prepared, the rule of 80/20 will prevail. Eighty percent of the talking is done by 20 percent of the people. That 20 percent is almost predestined. Any of the remaining 80 may intrude themselves into the discussion and their input will be welcomed. Comments will be made about how glad everyone is that the point was made. However, after about the third time, the intruder begins to get the message that 20 percent is 20 percent.

Meeting culture drives more talented people out of companies than any other hassle, in my opinion. They drive through three or four companies until they reach the level that dominates the meetings. Then they promptly forget what they have learned and begin to enjoy sessions for the first time.

There are dozens of similar activities conducted by companies, and none of them can be found in the policy book. I have met dozens of CEOs and never found one that set out to hassle the people. In fact, they all have a sincere set of comments on how their company is a "family" and how all the employees go out of their way to help others.

"In fact," one told me, "we even surprised the marketing department last week by painting their offices over the weekend. They told me they were delighted."

They said they were delighted, but were they? It will probably take six weeks to straighten out all the mess made by the painters who moved the furniture and papers around. No one told the people in the office it was going to happen, no one let them prepare, no one asked them what color they would like.

Being an employee in a hassling company is a lot like living at home after you grow up and having your parents decide all kinds of things for you.

Life has some built-in hassles that we bring down on ourselves. What we do to ourselves is our business, and we probably get what we deserve. However, there is no good reason why others should do unto us. Hassling can be prevented, not item by item, but by learning how to communicate both ways. Hassled people just do not produce quality work; sometimes they do little work at all.

A "hassle" company is one in which management and employees are not on the same side. The "hassle-free" company is one in which all employees are together and there are no sides.

It is possible to tell within fifteen minutes or so which kind of company you are visiting. "Hassle-free" offers pleasant working relationships, a smooth system, and happy employees. It also produces an environment for maximum profit and growth potential. Customers can identify this type of company and have confidence in it.

"Hassle" means that the people inside the company spend more time working on each other than they do making something happen. After all, most of the actions an individual takes during the day are instigated by something that is happening nearby. So arguments, checking, and disruption are inside jobs. It is said, for instance, that a corporate staff doesn't need any regional offices or factories to keep it busy. Its members can create work for each other. The irony of the hassle company is that it isn't something created on purpose or through malicious intent. It just seems to happen, and thus it can be avoided. Avoidance is primarily a matter of attitude and communication.

Of course, there are pathological situations that are not very fixable: the boss who is so concerned with being important that no human communication is possible; the devious or corrupt leader; the power struggle that becomes open warfare in which villages and villagers are destroyed. It is best to abandon those situations and search for a more fertile area. Fortunately, they are extremely rare.

The actions that hassle employees and create a negative atmosphere are usually not big items. Several of them are discussed in the chapter on demotivation. Here is the same situation as it occurred in two different companies. We can see how it was handled and what the ultimate result was.

## SITUATION

The operations department became concerned about productivity. A recent report indicated that 43 percent of all white-collar work was involved in the extra effort needed to get things done on time. The efficiency of the entire system seemed to be slipping each year. Random checks indicated that the supervisors were also concerned about productivity and the need for improving it.

## Company A:

Operations director: "George, I am concerned about productivity. I asked you to go to a couple of conferences on the subject and figure out what we should do. What is your conclusion?"

George: "I have studied the subject quite a bit. There is a lot of attention being given to it all across the nation. All the national figures show that productivity is low and getting lower. We need to take action quickly."

OD: "I agree. Let's get the drive moving."

George brought together a group of supervisors of clerical and assembly operations. He told them that productivity was low and that it needed to be improved.

The group felt that the problem lay in people's just not working hard enough and the company's not making the job practices clear. They decided on what had to be done, issued the orders, and then waited to hear what happened.

A few months later George reported that things were pretty

much the same except that the new productivity improvement program seemed to irritate people.

"They just don't understand that productivity is important," said the OD.

## Company B:

The operations director commissioned a study on the company's productivity and found that no one really understood what the word meant or whether the company was good or bad at it.

Key executives went out and obtained an education on the subject, learned how to measure status, and then made an evaluation. This evaluation showed that the company was taking longer to perform data processing and other communications tasks, and that there was a great deal of clerical rework due to procedures not being completed properly.

The action team brought representatives of each department together to form a communication base, got professional help in redesigning their data systems, and made it possible for all employees to contribute to the overall process.

Company B is doing just great. Company A is still worrying about it.

The process of running a hassle-free company involves a lot of actions and policies. Everything that is done has to be realistic, consistent, and imaginative.

If this were a class in teaching managers not to cause hassles, it would be set up in several sessions. There might be three of them.

## SESSION ONE: "THE REALITY OF HASSLE"

*Video:* Two salespeople are sitting at a coffee counter. One looks glum. Since there are very few people around it is obviously mid-morning. The glum one speaks.

"I get so discouraged sometimes. My brother-in-law is thinking about opening a sporting goods store and he has asked me to go in with him. I just might do that."

"But your sales records are good."

"Oh, sales are no problem, the customers are great. We have had a lot of product quality problems, but I can handle that. It's our office that bugs me. . . . They want more and more forms, they keep changing the product information, and they think we are all lazy louts."

"We used to have that problem, but we finally got most of it straight."

"How did you do that? Did everyone threaten to quit?"

"No, we had a session for the sales staff and the office staff. We took all the things we have to do for them and taped them on a wall."

(Scene shifts to the conference room with paper taped up.)

Leader: "Okay. Now we have put up all the paperwork. The object is to see who really needs what. The theme is: Realistic."

(Back to coffee house.)

The glum salesperson: "What happened?"

"Well, it turned out that about 80 percent of what was in those forms wasn't really used by anyone, so we got rid of it. And a system was set up whereby the office people go out and spend a day with the salespeople on a regular, planned basis. Then the salespeople go into the office periodically."

"What is going on now?"

"Sales are up. Hassle is way down, and I am able to handle a much larger territory. That has helped my commissions. Gotta go now. I have an appointment across the street."

*Workshop:* Divide the class into four groups, and ask them to identify any requirements they have to meet that cause hassle. These would include specific actions such as filling out forms, making reports, and such. Then the requirements should be marked to show where they originated and whether everyone participated in their development.

*Discussion:* Reassemble the class and ask each group to list their items. The purpose of the discussion period is to identify any items that cause lost time and hassle. The question should be asked: Are they really necessary and important? If not, why do we have them?

*Work assignment:* Go back and eliminate one useless bit of irritation and be prepared to report back to the class on the result.

## SESSION TWO: "THE VALUE OF CONSISTENCY"

*Video:* A baseball coach calls his team to the center of the diamond. He has a notebook and some procedural-looking papers.

"Okay, listen up. We have some very important information here that has come down from the front office. I want to make sure all of you understand it."

Player: "What's up, skipper? Are they moving the franchise again?"

"No, it's not that simple, Lefty. They have instituted a new program to increase attendance. The fans like to see more action than they have been getting around here."

"But we are leading the league. And I pitched a two-hitter yesterday. We're doing great."

"Two-hitters put people to sleep. What they want are a lot of hits and a high-scoring game. So there are going to be a few changes. Now the pitching mound will be moved back about 20 feet."

"But the pitch will be just coasting by the time it gets there."

Coach looks up and nods. "And first base will be moved 15 feet closer to home plate. That way the batter won't have so far to run."

"That will throw second base way out of line."

"There ain't going to be any second base. Third base will be second base. That will make it a shorter trip around the bases."

"You might as well bring the walls into where second base used to be."

Long pause as they look at each other and the coach turns the page. They look at the sheet and both nod accordingly.

*Workshop:* Divide the class into four teams, and ask them to identify some changes that were made and then unmade—which indicates that they were unnecessary in the first place.

*Discussion:* Review the items and determine why some of them were changed. Was it lack of information? Was it lack of concern for the effect?

*Work assignment:* Go back and identify an item that the employees feel has been changed for no good reason. Find out what part you played in making that happen.

## SESSION THREE: "THE NEED FOR IMAGINATION"

*Video:* Scene in the company cafeteria. Two people are standing, looking at the food display.

"Do you realize that they have exactly the same menu each

week? And the salads are the same every day. I doze off while I'm eating."

"Let's drive down to the fast-food place."

"It's hard to get in there at noon time. No one eats here any more."

"But actually the food isn't so bad."

"No, it really isn't, but it's so mechanical and dull around here. It's like they don't really think anyone eats the stuff. They have an obligation to put it out and bring it back. Do you think they even see us?"

"Put a blindfold on and let's go through the line."

*Workshop:* Assign speech topics to the attendees, each speech to be impromptu and cover no more than two minutes. Ask attendees to list the most important aspect of making a speech. After all the responses have been listed, remind the group that the speaker has an obligation to be interesting. There are no dull subjects, just uninteresting speakers. Ask some people to give an interesting speech—showing imagination.

*Discussion:* Why is imagination an important part of the management communication system? What dull things could be improved here that would help dehassle the place?

*Work assignment:* Go back and take a look around. Determine what job or place your coworkers dislike the most. See what it would take to reverse that feeling.

In a hassle-free company the employees have confidence that the management respects them and needs their output. They know that the requirements of the job are clearly stated, and they have had the opportunity to make inputs to that statement.

They recognize that management is committed to performing to those requirements and takes them seriously. They see that recognition is accorded to those who do well and help is given to those employees having difficulty. They see that management has its prerogative but for the most part shares the load. They respect management.

Making all this come about requires that the company be involved in a lifelong process of quality improvement. This process touches each employee, each function, and each bit of output, whether it be service or product.

The next few chapters concentrate on making quality happen. But the whole dehassling business is a process, not a program.

Eating a sandwich is a program; raising children is a process. You never get done with a process.

The value of the hassle-free company is obvious for the most part. One aspect that may not leap immediately to mind is that it is a great place to look like a sensational manager. When all you have to do is the job, there is plenty of time to cause a lot of good things to happen.


## THE PLANNING HASSLE

Wanting to make certain I offered a detailed example of corporate planning, I searched around to find the company that other planners thought had the best system. I anticipated that this would be a difficult task but quickly learned that the "pro's pro" of planning was the Lightblue Corporation. Its system was revered in planning circles, and CEOs were constantly hiring LB's junior people to set up similar systems for their corporations.

There were no negative stories about LB or its planning whiz, senior vice president Harrison Wilson. Mr. Wilson agreed to see me in ten days and to devote a full afternoon to my questions. Then he promised to set me up with other executives who used the system in their day-to-day work.

True to his word, Harrison Wilson greeted me at 12:30 and invited me to join him in his office.

"I like to start meetings at this time," he said, "because everyone is at lunch and no one interrupts me. I only eat an apple myself so lunch is not a big deal."

We exchanged pleasantries for a few moments, discovering that we had several mutual friends and that we may even have played on opposite sides of a Connecticut intraclub golf match the previous year.

Harrison asked me how much I knew about the LB system. I replied that although I had some information, it was sketchy at best, and I would appreciate it if we could just start from the beginning as though I knew nothing. This seemed to please him, since he commented that it made the whole process much easier.

"What we are after in our system is to know exactly where we are at all times and to be aware of just what is driving our various operations. Mr. Rocque, our CEO, is very determined that we will get no surprises. But he is just as determined that the executives we assign to run operations have enough freedom of movement to take action when it is necessary. So we are looking for an optimum. In fact I call our system the 'optimal action' concept, OA for short."

I noted that LB was a diversified company with six groups, of which four were manufacturing and two were financial or service-oriented. This much I had determined from the reading material Harrison had sent me.

"Yes," he said, "we are diversified, but we use the same basic OA system in all divisions. What is important is the activities of the people involved, how they run their responsibilities, not how the products or services themselves are viewed. That is part of the marketing concept, which we integrate into OA, of course."

"So your planning is of and for the people," I smiled. He returned the smile rather wanly but nodded.

"There is certainly no reason for doing it at all if it doesn't benefit anyone. That is one reason I hold formal reviews each quarter just to see whether the material is being used. We have no allegiance to planning for the sake of planning. It has to stand the test of usefulness.

"In the spring, we cover the market projections for the following three years—what do we see happening, how will it affect each division, and what is the overall outlook. The marketing people lead this but my staff pulls it all together.

"Each plant of each division comes in with its projection; at that time they all get to hear what is going on in the rest of the corporation since everyone attends the basic presentations. It takes about two weeks to go over everything."

"Where do you do this?" I asked.

"We usually take over one of the big meeting sites in the foothills of the Poconos," he replied. "They give us a good rate since nothing much is going on out there then. And one of the real values of this type of meeting is the chance all the executives have to mingle and get to know one another.

"After the marketing survey is complete in early summer, we start right in on the financial projections. These are done at the group and division levels. We want to know what the sales, margins, inventory, compensation, and other figures are going to be for the next three years. We also ask for the fifth year but that is just for exercise."

"Who puts this together from the plant standpoint?"

"The general manager and the department heads primarily. We have a planning support activity in each plant representing my office but the main work is up to the plant staff. We want it to be their program. Otherwise it just won't happen."

"When does all this financial projection information come together?" I asked. "Do each of the plant managers send it right in to your office?"

He shook his head patiently. He wanted me to really understand how this system worked.

"The division presidents pull it together and then work with their group executive in order to ensure that it all fits. Usually the groupie will bring it in to me or perhaps discuss it with Mr. Rocque to see whether it all tracks.

"Then we put all the figures together and check it out with the comptroller, who relates the margins to profit and balance sheet figures. When all of that is ready, we have a full meeting, usually around the Fourth of July weekend, and everyone gets to see the numbers at the same time."

"Where do you have that meeting?"

"Right here in headquarters. We have a room on the top floor that can seat 175 comfortably and still let them see the screens. I'll make sure you get a tour of our room and its projection setup. We have something that is unique in corporations.

"Mr. Rocque feels that people work harder and have more loyalty if they can have more information. That is one reason that we concentrate on that aspect."

"And how long do these meetings last? Is everyone in attendance at all the meetings?"

"The overview, presented by my staff executives, lasts about one and a half days. Then the division presentations take the rest of the week. However, we don't ask division personnel to stay for sessions by other divisions unless they want to. I'm pleased to say that most of them seem to want to. They get very involved."

"So you now have the market projected and you have the financial aspects well understood and stated. Is the rest of the time spent in tracking these objectives?"

"Oh my, no!" he exclaimed. "This is just the beginning. It doesn't do you much good just to know what you want to achieve; you have to get down into the trenches and determine exactly what is going to have to be done to meet those objectives. As soon as the financial planning is over, we start business plans.

"Now business planning is where it is all at. Each operation has to say what they are going to do to meet these objectives, how much capital, if any, they need, what kind of problems they visualize, and what they think the competition is doing."

"How long do they get to prepare these plans?" I asked. "Do they send them into your office for review?"

"They come to our office but only after the division staff, the group people, and headquarters staffs have blessed them. We used to have them come in here but they were really too incomplete. That is one reason we set up our planning school. Every plant department head in the corporation goes to that one-week session. Plant GMs and up attend a special two-day session."

"Every year?"

"Yes, every year. We want to keep them up to date on the best techniques. And we usually have the two-day session following the president's state-of-the-company weekend. The group executives and division presidents attend that along with the senior corporate staff. We just hold a few over and do the course. They love the chance to get together."

"Where do you conduct your planning school?"

"We do it in different nice resort places like Greenbrier, Broadmoor, The Boca Raton Club, and others. It gives them the feeling that we are serious when we do it that way. I make a point of dropping in on each class to show the flag."

"Are your executives pretty dedicated to planning on this scale? Doesn't it take a lot of their time?"

"It doesn't take as much time as you might think," he said. "But it is an investment by the corporation. We think it is a good one and that we would be derelict if we didn't do it. We can pinpoint all our activities right down to the last and smallest item. If something goes wrong, we can determine exactly why and learn from it.

"But where we really pull it together is in the business plan presentations. That is the most exciting part of the year. Each division tells us plant by plant, operation by operation, what it is going to do during the coming three years and how it performed against the previous three years."

"When do you do these sessions, and where?"

"All these, except for the ones from the Latin American and Far

Eastern operations, are done right here. We go to São Paulo and Hong Kong for the others. We schedule most of November and December for the BP meetings. All the top management sits through every one.

"The old man makes certain that he asks each manager at least one question; that way they know for sure he is interested in the outcome. He really is a master at it, and he never forgets what you have promised him."

"After the business plan meetings, what happens to the plans?"

"Well, the actual plans themselves are stored, but the operations staff begin immediately auditing the operations to ensure that they are meeting or exceeding their programs. When a problem arises, a special task team is formed to help the operation overcome its difficulty. Since we know what is supposed to be happening, we can tell immediately when something is going astray."

"And after the first of the year . . . ?"

"We begin the planning process again. It is a continual process because we live in an ever-changing world. I like to think that ours are living plans; we can always be up to date and can hit the ground running when we need action."

"Have you figured out what this process costs?"

"Not really. We are more concerned with what it would cost if we didn't do it. There's no telling how we would operate. But the most direct cost is the planning staff—about ninety-seven people throughout the company."

Walter Thomas, group executive, was an unqualified supporter of the OA system. "Five years ago we never knew where we were going or even where we'd been. Now we can make a clear plan and track it week by week. I meet with each division president twice a month and get them all together every six weeks."

"Is that your main status meeting?"

"Well, we have the monthly corporate meeting with the CEO and his staff. You'd like to see that one; it is really a well laid out affair. The comptroller gives the status, the headquarters department heads talk about their function, the treasurer displays the accounts receivable, manufacturing gives an inventory presentation. Then each division president makes a statement and is quizzed on the operation. It is a really meaningful session."

"How long does it take?" I asked.

"Usually it is only one and a half days but once in a while it goes on through the second day. If something significant comes up, Dave might ask for a task team to meet the following day to begin a corrective action program.

"It really helps a division president to be able to call upon a team of people from finance, manufacturing, training, quality, engineering, and other functions quickly. Sometimes they don't want to move as fast as corporate does, but we are learning to work it out."

Thomas took me to lunch in the executive dining room where we sat with two of the functional vice presidents: Howard Gibbons, manufacturing, and Carl Watson, quality. They good-humoredly told me tales about the monthly corporate meeting, mostly revolving around the trials and tribulations of those who hadn't been able to meet their commitments.

"One of the problems with all this planning is that everyone winds up with obligations that hang right out there in plain view. If you don't bill enough sales, or ship enough goods, or sell the right number of policies, you are in deep trouble. After all, you are the one who constructed the plan in the first place."

Watson reached for a celery stick. "I agree with Howard," he said. "Measurements are there for everyone and are quite visible. Sometimes I think that it tends to restrict us, to limit our stretching out to go after something that we're not sure we can accomplish."

Walter Thomas leaned back and frowned at this. "I don't see anyone holding back at all," he said. "Sometimes I have to pull them back a bit to something I think is more realistic. We have to have plans that reflect real life."

"In real life," noted Howard, "things don't come to a halt while you prepare yourself for battle. We have each thing laid out so specifically before we begin that once in a while the whole game is over before we even get out of the locker room."

"I know we've had a little of that," said Walter, "but I think the trade-offs are clear. We had some real disasters in the past, you'll have to admit that."

"No question of that at all, we've probably avoided some real clunkers. But just the same I'm glad I'm here in headquarters running a function instead of back out there running a division. Here we can bury our mistakes and no one ever knows."

How did they feel about the annual objective-setting meeting held in the foothills of the Poconos, I asked. There was a brief silence, and then Carl answered. "The only thing worse than going is not being invited to go."

Deborah F. Greese, division president for insurance and finance, was my next appointment. Greese had come to LB along with her company several years before and had been rapidly promoted until now she ran the division that produced 20 percent of LB's sales and 40 percent of the profits. After assuring herself that I was not writing for a current business magazine but was merely doing research for a book, she decided to open up.

"Frankly," she said, "I think that we overdo this whole thing. If I attended every planning session I am supposed to, plus the status reviews with the group executives, the monthly corporate meeting, the off-site training sessions, and the division presidents' meeting—well, I would have about three days left in the year to run my companies. And don't forget that the people who work for me have all that to put up with plus meeting with me, so they must be working on a deficit when it comes to time.

"Add the traveling necessary to know about your operation, and you will find that this thing is impossible. Now, it wouldn't be so bad if anyone used the information besides headquarters. As far as I can tell no one does; at least my folks don't. They have their own tip sheets. We have a separate, two-sheet activity listing that we use for our meetings. None of the forms or books or other planning things that Harrison Wilson puts out exists anywhere except in his office and maybe the CEO's bookcase."

I was thunderstruck. "Are you telling me that the whole exercise is a big useless hassle? Do you mean that all the hundreds, thousands of hours of executive time are just being frittered away?"

"Well, not all of them. We probably wouldn't ever see each other if the meetings weren't held. But we could accomplish the same thing with bimonthly two-day meetings. One planner per division could pick up the plant-by-plant or, in my case, product-line-by-product-line plan and put them all together, and then if the old man and Harrison didn't like something, they could bring it up in the meeting. You really don't need more than that."

"Then why do they do it? What is the purpose?"

"Purpose? The purpose is to create an image that Lightblue is a magnificently managed corporation, which will then result in the share-price-to-earnings ratio rising. That will make all the management look good and we'll become famous."

"It must be working; your corporate growth looks pretty steady."

"Most of that growth has come from the investments in the insurance portfolio and growth in the electronic components divisions. They both grow or shrink independently of any planning process. In about one year, just about the time your book will come out, we should reverse both those trends. Right now we need to get rid of several operations and diversify into other areas. However, it is hard to make a plan that will pass the test of being absolutely infallible."

David Rocque, the chief executive, chairman, and president, was a delight. Despite his list of titles, he insisted on being "Dave" to me and to his employees. He disarmed me immediately and told me that since he didn't have a great deal of time, why didn't I get right at the meat of what I wanted to ask him instead of taking time to fool around? So naturally I asked him how much he depended on the elaborate planning system, and all the status reporting that went with it.

"Whenever I want to know something, I just call up the person in charge and ask him or her what's going on. They always tell me. And of course the comptroller and other functional directors keep me up to date, with particular emphasis if there is a problem. So I think I've got a pretty good view.

"All the planning and formal status reviews help get the operating people to think about their work in a planned fashion, perhaps leading them to do some things they wouldn't do ordinarily. They all seem to think it's valuable so I don't touch it."

"You mean that the planning system isn't for you?" I gasped. "Who is it for, if it isn't for you?"

"Why, it's to run the company. Just because I don't use it as my primary source doesn't mean that it isn't the keystone of our system. You'll have to get Harrison to give you a rundown on it. He really is a brilliant young executive."

That was as far as we got. Shortly after that, I left the Lightblue building to go back to my office and sort out my notes. As near as I could determine, the only ones who wanted anything to do with the planning and control system were those who administered it, a few old-timers who thought the boss really wanted it, and the owner of an inn in the Poconos.

| Item | Travel Days | Meeting Days | |
|---|---|---|---|
| Quarterly Planning Review | 2 | 3 | |
| Spring Marketing Projections | | | |
| In Plant | | 15 | |
| Division | 2 | 4 | |
| Group | 2 | 3 | |
| Financial Projections | | | |
| In Plant | | 15 | |
| Division | 2 | 6 | |
| Group | 2 | 3 | |
| Corporate | 2 | 3 | |
| Business Plans | | | |
| In Plant | | 20 | |
| Division | 2 | 3 | |
| Group | 2 | 3 | |
| Corporate | 4 | 10 | |
| Training | | | |
| Senior Execs | | 2 | |
| Others | 2 | 5 | |
| Status Reviews | | | |
| Division (quarterly) | 8 | 8 | |
| Group (monthly) | 24 | 24 | |
| Corporate | 24 | 36 | |
| Totals | 78 | 163 | |
| Executive Working Days | | | 242 |
| Executive Planning Days | | | 241 |
| Remainder for Running Operations | | | 1 |

This entertainment system was costing the company at least $45 million a year in employees, support equipment, and most of all executive time. Further interviews with plant and office general managers showed that each had to have an assistant to run the operation while the manager concentrated on keeping the headquarters people happy. The top five levels of the company didn't do any fruitful work. They were—to use an old middle-Ohio expression—"taking in each other's washing."

Consider what a typical operating executive invests in time for this system. Each discussion had to be held on several levels with the executive participating each time.

To me the best evaluations are unscientific. "How do you feel?" for instance, brings forth a lot of unproven opinions. "What did you

| Task: GETTING A | Rating (0−10) |
|---|---|
| Parking space at work | |
| New desk | |
| Little time with the boss | |
| Change in insurance form | |
| Travel advance | |
| New mail number | |
| Raise for a subordinate | |
| New supplier | |
| Performance review | |
| Transfer | |
| Seat at the meetings | |
| Little peace and quiet | |
| Vacation change | |
| New product approved | |
| Policy issued | |
| Total | |

| | |
|---|---|
| 0−37.5 | This is a good place to work. It probably does not have many staff people. |
| 38−75 | Most things can be handled, and this probably is a reasonable group of people. I suspect people do not realize the amount of hassle being generated and will be willing to change things if asked. |
| 75−112.5 | Here we are beginning to deal with the professional harassers. They are not going to change. Start looking for a way out. |
| 113−150 | Don't even finish the day. Escape while you still have a steady hand—unless you enjoy being hassled and hassling in return. There are those who fit that description. They complain almost all the time about the company and the terrible, insensitive things it does. However, while they are chanting their litany, they are producing an even greater hassle for the listener. |

think of the show?" "Is their quality any good?" "Where is a good place to live?"—these and other questions get answered somehow and communication is made.

Hassle means different things to different people. But to me it is the unnecessary difficulties or harassment placed in the path of someone trying to do a reasonable thing. It is what people mean when they talk about the effect of bureaucracy, nonpurposeful actions that made things difficult.

Knowing all this, I have created the Hassle Index. It provides an idea of what the hassle level is in a company. All that has to be done is to determine how much trouble it is to accomplish some perfectly reasonable task. The hassle for each task is graded from 0 (no sweat) to 10 (the IRS is a pushover by comparison).

There are fifteen items listed on page 34. I'm sure that you can think of dozens that are more applicable to the business you know best. However, it is possible to have a score of 150, in which case the evaluator can be considered to be working for the devil, or zero, in which case there is some obvious dishonesty at play.

# 4
# A Quality Carol

Emory Spellman prided himself on being the first in the office every day and the last out at night. Ambitious young executives soon gave up trying to impress the president by working longer hours. They learned that Spellman considered them only a little less weak and unresolved than the others—those who came and went at the regular times.

Spellman had run Consumer Consumptions (CC) for thirty-five years. He and his friend, Jacob Masters, began with a variety store. Over the years it had grown into several retail stores, kitchen appliance manufacturing and service, a credit card system, consumer financing, and even an automobile rental company.

CC had always been profitable because Emory and Jacob made certain that their prices were below everyone else's, except when it came to the financing operations where the interest charged was as high as possible.

In the past ten years CC was losing market share and profitability because the customers were not satisfied with CC quality. This was particularly true in the credit card operation, where 65 percent of the accounts typically had at least one mistake every month. The refrigerator and dishwasher factories were having to deal with recalls and high warranty. The CC field service people were considered the best in the business. However, even they were having difficulty keeping up with the problems of CC hardware and services.

Spellman blamed the quality problems on the falling standards of the workers and ordered disciplinary actions. His office assistant, Walter Compton, often said, "Mr. Spellman thinks that if we could return to the old days, everything would be all right again."

Spellman was a fit man, strong physically, and ahead of everyone else mentally. He marched through the halls of the company with a firm stride and a firm destination in mind. When he traveled to a CC facility, it was well known that he could find inefficiencies even the most experienced consultant couldn't unearth.

Jacob Masters ran the manufacturing operations, and Spellman managed the service and financial functions. Spellman was the senior executive. Masters' last management meeting brought out some typical CC problems. Since Jacob went home from this meeting and quietly died, Spellman was left to resolve the problems that Masters had raised.

"The refrigerators are having four service calls during their first year," reported Masters. "We only plan for 2.45. This higher call rate just about eliminates any profit on the boxes."

"The service people are not finding all the problems during the first call," said Spellman. "Get some new service people and get rid of the old ones."

"I think that will fix it," said Jacob.

"Production is down in the dishwasher line," said Masters. "They stopped the line several times last week in order to fix something."

"Stopped the line? Stopped the line?" roared Spellman. "Are there no rework areas? What kind of place are we running around here? If we stop every time some little thing goes wrong, we won't have any product to sell. That is what we have field people for."

"The credit card people say that they need more training programs in order to help people keep from making mistakes."

"Tell them to hire trained people. We're not a university. We can't coddle employees, they want too much as it is. Do more inspection of the bills before they go out and fire those who can't make out a statement right.

"Last week someone came to me and wanted to hold up the direct-mail advertising brochure because there were a couple of little errors in it. I told them that quality was a trade-off and that the people who would pick those things apart weren't going to buy anything anyway. You people are going to have to quit trying to gold-plate everything around here. We can't afford to spend money on quality."

A week after Jacob's funeral Spellman took the late bus home, as usual. When he reached his stop it was dark, as usual, and the streetlight was not working, as usual. He stepped heavily down from the bus, papers under one arm and briefcases in the other.

As the bus pulled away, he noticed that the place seemed a little different. In fact it was a lot different. This was not his stop at all; he had gotten off in the wrong place. There was a factory building where nothing had stood before. A door, with a lighted room behind, was the only thing he could make out. It was a most uncomfortable and frightening feeling.

Spellman looked about but the road itself seemed to be gone, covered by a rolling haze. Somehow it was apparent that no more buses were going to be traveling through there in the near future.

Recognizing this, Emory advanced with hesitant steps toward the lighted door. In there was a telephone, certainly. He could call a taxi, or perhaps his neighbor would drive down to get him—if he could determine where this place was.

An older and very tired woman was sitting at the receptionist desk and looked up as Emory entered the room. "You must be Mr. Spellman," she said.

"How could you possibly know that?" he stammered, sensing a rush of lightness in his head and feeling certain that he had turned pale.

She did not seem to notice.

"It is time for your appointment. Go through that door over there, turn right, and walk down to the fourth bin on the left. He's expecting you."

"Who's expecting me?"

"You're wasting time, get moving. Fourth bin on the left."

She turned back to her magazine.

Angry by now, Spellman had forgotten his fear. For no particular reason he suddenly felt the same bitterness at employee incompetence that had come over him when Compton had asked for the day off to go to a quality improvement seminar. He glared at her and then went through the door. The door opened into a large open factory, dimly lighted and dingy. Spellman was not a fanatic about housekeeping, but he felt that this place could do with a bit of cleaning. It was cold and depressing.

As he walked down the aisle, he passed the first bin and saw a man sitting in an office working at a desk. Papers were piled all over the desk, on the chairs, on the floor, on top of the many file cabinets. Emory stood there for a moment watching the man frantically working, but the man never looked up.

In the next bin a woman was moving suitcases of all descriptions and sizes about. They kept arriving through the far door, and she would first take the tag off each and put another one on. Then she would move the cases to another place and shove them through the door, but they would come back in the first door.

Confused, Spellman proceeded past the third bin, noticing only that it was empty. He stopped, feeling a flush of fear that quickly passed as he reasserted himself and stepped in front of the fourth bin.

It was here that he almost lost it all, because sitting at a bench was his old friend Jacob Masters surrounded by work. There were refrigerators, toasters, dishwashers, motors, stoves, washers, dryers, electric razors, and all by the dozens. There was hardly enough room for Jacob and his toolbox.

Jacob looked up only briefly. "Yes, it's me, Emory. I can't stop to talk very long; we have to get this over with quickly."

"What is going on here, Jacob?" shouted Spellman. "What is this place? You have a very rude woman out there in front."

Jacob looked up over his glasses. "Shut up and sit down, Emory. I had to promise them many things so you could come here, so don't waste time yelling at me. I wanted to give you a chance to save yourself. Before we start talking, take another look at that bin next to me, the third one that is empty. And be quick about it."

Spellman rose quickly and ran to the edge of the bin to look in next door. What he saw almost bowled him over: his office was in that room. It was piled with paper, and behind it the wall had opened. He could see vast stores of products, the same kind as his friend was now working on.

He rushed back to the room. "What is this, Jacob? Please tell me. What are you doing? Why is my office next door? What is all this?"

"I am fixing all the problems that I caused to be put into products and services and every other aspect of business for the past forty years. Do you remember back in 1957 when I told them to go ahead and deliver the new washing machines, even though some of the holes were a little oversized?"

"Vaguely," said Spellman. "We always made a lot of those decisions."

"Well, we had 24,871 machines that leaked. All of them are sitting back there and I have to fix them. See these toasters? I bought wire for them that didn't meet the specifications and some caught fire. I have to rewire all 367,298 that we made."

"What are you being punished for?"

"For being the cause of the hassle other people had to live with. For not preventing these things by being interested in quality— that's what I'm being punished for. And I am going to sit here forever, day after day, until the end of time, because when I finish all the products, I have to start on the service company's paperwork. That alone will take forever."

"And my office next door?"

"They are getting it ready for you to occupy some day. Oh, you'll get to live your full life out. These folks are in no hurry, they have forever."

Jacob sighed and went back to work on the toaster as Emory absorbed all this. He shook himself, and then gave the back of his hand a pinch as reassurance that it was all real.

Jacob spoke again. "I have this computer terminal here to keep track of where I am at. They are very thorough. Every last, stupid, short-ranged decision against quality I ever made is in here. And the total effect is considered. Remember when I cut out quality control in the foundry?"

"Yes, that saved us $35,000 a year."

"Well, because of that, there are seven—" he checked the computer "—no, there are eight railroad cars full of castings out there waiting for my attention. And if I don't make the output requirements every day, they add to my pile."

"What do you want from me, Jacob?" said Spellman. "What does this have to do with me? Why have you brought me here?" Jacob started to answer but Emory interrupted. "What is that woman doing with suitcases? They keep going around in circles."

"She was in charge of luggage control for a big airline. Those are all the lost suitcases that never found their owners because she wouldn't train her people or have procedures that could be understood. She will never get done.

"And neither will you, Emory. All these years you have treated quality like it was something you could put in or take out. Well, unless you change your ways, you are going to wind up right next to me forever and ever, twenty-four hours a day. No time off, no visitors, no meetings—just all the problems you ever caused."

"That's a terrible thought, Jacob."

"Tell me about it," said Jacob.

"But I'm only doing what everyone else does. I'm not an evil man. I don't want to hurt anyone. My responsibility to the company is to make money and to be efficient."

"I think you are going to find that the way to make money is to give the customers what they were promised and eliminate the hassle for the employees. At any rate, you have a chance. I suggest you take it."

"What do you mean, I'm *going* to find?"

"You are going to be visited by three teachers: quality past, quality present, and quality future. What you are seeing right now is your future if you don't do something about it. I have to go now, Emory, and you have to get out of here."

Suddenly, everything disappeared, and Spellman heard a voice saying, "You have to get off now, Mr. Spellman."

He realized that it was the bus driver speaking to him and that he had dozed off. Hesitantly at first, and then quickly, he went down the aisle and out the door, relieved to see familiar surroundings. Going down the street to his home he quickly went in the front door and sat down, packages and all, in the chair.

"Don't get too comfortable. We have to be going right away," said a voice.

Spellman leaped from the chair and looked across the room. There he saw an old man clad in an academic cap and gown. The professor beckoned to him. "Do you remember me, Emory?"

"Professor Barrington. Certainly I remember you, sir. I had several classes with you when I went to college. You were always very interesting."

"Thank you. Alas, I was also somewhat misinformed. My assignment at this time is to try and undo some of the problems I may have caused by teaching my students what turned out to be impractical things. Where you are concerned, the subject is quality."

"Impractical? But we always concentrated on the practical. I thought the idea was to cause the most efficiency. And certainly quality is part of that. We spend a lot of time deciding how good things have to be."

"Let's walk through this door and take a little trip into the past while we are chatting."

Taking Emory by the elbow, Barrington guided him through the dining room door and into a classroom scene where attentive students were listening to a younger version of the professor:

"We must identify the characteristics of quality that are the most important so people can know which ones have to be met exactly. This is the efficient way because there is an economics of quality. If you have too much quality it costs money, so we must find just the right level."

Spellman turned to him in agreement. "That is exactly right, and that is the way Jacob and I always operated. We learned to make judgments all the time. It worked out well."

"Well, it sounds good, but the problem is that this approach is just too fuzzy. It meant that the talent of the company spent all its time deciding whether things were good enough or not.

"As long as we had cheap energy, low labor rates, a captive market, and inexpensive material, we could get away with that. But when everything had to count, it just became impossible to be so wishy-washy."

"I don't understand what you are saying. Quality isn't a specific; it isn't something you can be exact about. You either know it when you see it, or you don't. Look at those fellows over there."

The two approached a wire-caged area marked "Material Review." Inside, two young men were discussing a part lying on the table.

"I know the hole is a little oversized, but we can add some sealant to the rubber ring and it will work perfectly."

"But, Jacob, if we do that and the sealant dries out later, the machine will leak. I think we should go back and make them over. What do you think, Emory?" They turned to Spellman and the professor.

Emory hesitated, then said, "I think Jacob is right. We can't get every little thing right or we will never ship any product. I agree on fixing it up with sealant." Emory turned to the spirit. "I didn't think they would be able to see us."

They see you, but not me. They see you as you were. That is the part Jacob is working on right now back in that awful place."

Emory shuddered.

"It's a real mess. You see, the part was designed to fit with other parts. When it does not conform to its own requirements then the other ones won't come together. That sounds obvious, but it is true.

"You and Jacob, and the other people of your time, learned that it wasn't necessary to take the requirements seriously. You felt you could compensate for anything that went wrong."

The general manager, Al Spangler, walked up to Emory. He was shaking his head in concern.

"We are going to have to be more careful about quality from now on, Emory. I can tell you that there is a lot going on and many people are worried about it. Last week the quality manager and I were invited to a meeting with one of our big customers. They are on a quality kick and wanted us to know that we are going to have to do better. They aren't fooling around.

"They were talking about zero defects—material that conformed to all the requirements all the time. Then they wouldn't have to have any inspection or test at all." Spellman winced.

"That would cost a fortune. We'd have to do ten times as much checking to see that no defects happen. It would be impossible. Let's kill this one in the bud. Just stall around and they'll get over it."

Al shook his head. "They expect us to give them a plan as to how we are going to start preventing problems. They don't want us to inspect more; they want us to prevent more. They are really serious."

Spellman pondered this a bit. He wanted to ask the professor's opinion but was afraid that Al might think he was talking to himself.

"I'll tell you what we can do. Let's get all the workers together, tell them that from now on we are all going to do everything right the first time, invite the customer to speak to them, have a band—and let them think we are going to do something different. Get together with Quality on it and move out."

"That's a great idea, chief. I'll start on it right away."

The professor shook his head. "Why do you think it would be so impossible to prevent problems? You have some very good people in this company. They spend half their time fixing and almost none in preventing. If you haven't tried it—don't knock it."

"But we are doing very well in our businesses. The TV operations make more from the service contracts than they do from the sets. The overhead in the financial operations is getting a little high, but one day we'll have better computers and they'll get the errors out. The car rental company is growing—we added ten more cities last month."

"Well," said the professor, "it seems to me that quality is more of a factor in all your companies than you think it is. And I hope you are becoming aware that some of the conventional thinking you developed over the years is quite old-fashioned. Even more than that, it is impractical, inefficient, and very expensive.

"I have to leave you now. You will have another visitor to discuss quality present. Goodbye."

Before he could think about it, Spellman was back in his office. "Now I have to take that bus ride all over again," he thought.

A young lady, "brisk of step and fair of face," approached him.

"Good morning, Mr. Spellman. I am Becky Thompson, your guide to review the status of quality present. I have been doing some studies of what is happening in today's world. Perhaps we could spend a little time together?"

"That would certainly be a pleasure, Ms. Thompson. But don't you think you could accomplish more if you talked to our quality control people?"

Becky smiled. "Causing quality to happen is beyond the reach of the quality control people unless management completely understands their role. It must be recognized that senior management is 100 percent responsible for the problems with quality—and their continuance.

"I'm afraid you are going to have to deal with me, Mr. Spellman."

Emory smiled and indicated resignation. "Okay. Where do we begin?"

Becky took a leather notebook from her briefcase. She opened it and removed a page, which she handed to Spellman.

"This will be our agenda. I thought we should first look at what our customers think of our quality, and then at what is being done to improve it."

"I have always been very proud of our quality," said Spellman. "We have had some specific difficulties, but on the whole we do very well."

"Okay," nodded Becky. "Let's begin right there. One primary measurement of quality is the price of nonconformance. That includes all the expense of doing things wrong. Wrong in manufacturing and wrong in administration, service, and other areas. Your staff does not know how much that is, but we took the liberty of having it calculated. Your comptroller has blessed it."

"Well, if I have to guess, I would say that it is probably 2, perhaps 3 percent of sales. That might be a little high, but I believe in being conservative."

Becky removed a chart from her briefcase and handed it to him. "We find it to be 23 percent of sales. Now that doesn't include the price of conformance—what is spent on inspection, testing, and education. Your company spent nothing on quality education, but appraisal costs are just about 4 percent of sales. So all together the cost of quality is 27 percent of sales. I don't have to tell you that it is nine times your after-tax profit."

Spellman fell back into his chair, motioned the young executive to sit down, and tried to compose himself. This was ridiculous. How could it be that he had never heard of such a collection of numbers? He managed to make a reply.

"I suspect you are going to tell me that if we could learn to do everything right the first time we could cut that number way back. How far back? Quality does cost money."

She sighed. "Unquality costs money. Fixing, correcting, scrambling—all those things cost money. You have already planned for the costs of doing something once. If we conform to those requirements, then there is no additional expense.

"But to answer the question, we see companies with a product and service mix similar to yours that have eliminated well over half the cost of quality in a year. We have one case of a company that has been doing it for some time whose total cost of quality is under 5 percent of sales.

"So, if CC really worked at it, after you understood it, this time next year it would be about 13 percent of sales. In three years about 6 percent."

All this was beginning to be a bit much for Emory Spellman. It had been a very long day. In fact, he wasn't certain that it had been a day. He had gone backward through time to when he was general manager—fifteen years ago. And now he was back in the present with this determined and incredibly efficient young lady.

He knew that he had to turn this situation around somehow.

"I don't understand all this fuss about quality. It has always been around; it has always been handled one way or another. Are you saying that there has been a decline in quality lately? If you are, I certainly agree that the workers of today are not as good as the ones of yesterday."

"Really, Mr. Spellman, you are making my job very difficult. It is not the workers who are any worse. For that matter, it isn't even management. You have to recognize that we never did make or administer or service anything too terrifically. But there was no competition except ourselves. It has only been since people from other countries began to send products in here that we could see the difference."

Spellman bristled. "I think you are wrong about that. Our quality is just as good as anyone else's and, as a matter of fact, we are doing a lot of things to make it even better."

He shuffled through papers on his desk and came up with the quality department report. With regained confidence he tapped the desk and addressed his interviewer.

"Here is one thing we are doing. We have ten quality circles set up in the electronic assembly plant. They have six members each and they have come up with some very good ideas for quality improvement. I certainly think that shows we are working on quality. Those circles, with varying membership, have been going off and on for two years."

Becky noted this in her folder. "How many people are there in CC, Mr. Spellman?" she asked.

"I suspect you know as well as I, but there are 3,500 at present."

"So the sixty people in the quality circles program are certainly a valuable asset. But that hardly represents any kind of commitment to improvement. Also we have done some checking and find that it has been very difficult to keep these groups going. The individuals feel they are all alone, that management is not part of it at all."

Spellman was concerned. "How can they feel that way? We are very supportive."

Becky leaned forward. "Mr. Spellman," she said, "you are going to have to stop deluding yourself about quality. CC has a growing reputation for shoddy service and merchandise. The management culture is very clearly oriented toward schedule and cost, with quality a distant third. You apparently do not believe this, but it is true nevertheless. I suggest you call your senior executives and ask them to name their biggest problem, and then ask them to tell you what they are spending the majority of their time working on."

Emory thought about this. "If I do that, and you determine that they are really serious about quality, then will you people leave me alone?"

"We are going to leave you alone anyway, Mr. Spellman. We are merely trying to shed a little light on the situation. The question is not whether you can convince me that everything is rosy. It is whether your company and all its jobs will survive with things as they are."

Emory invited five key executives to join the meeting. Becky moved to a corner of the room, thus eliminating the problem of having someone sit on her.

"I asked you to come together for a few moments so I can get a quick update on some things. What I would like to know is: What is your biggest problem? Then, what are you spending the majority of your time doing?

"The purpose of this is to let me get some background so I can determine whether we are going in the right direction. As you know, the business world is not all that smooth right now, and we need to be certain that we have the most practical strategy that can exist."

The five executives glanced at each other and then at Emory. After an initial moment of uncertainty they began to speak:

Helen Douglas, materials: "My biggest problem is suppliers that give us services, and products also, that just are not right. I spend most of my time with the lawyers and contract people trying to get all this fairly adjusted. And it is getting worse."

Bjorn Anderson, production: "Meeting the schedule is my biggest problem. We are having to make five of everything to deliver four. The work we are seeing today, combined with the materials we receive, makes it incredibly difficult. I am spending the majority of my time with the budget people trying to find a way to keep the production effort profitable."

Harrison Ellis, quality: "Our inspection effort is spread very thin since we have to do a lot more checking now. And we are beginning to work with the retail stores and the credit card operations to help them find problems earlier. I have been spending most of my time training inspectors."

Bill Davis, field service: "My biggest problem is keeping our products up and operating in the field. The number of service calls has doubled in the past year. I think that is because the customers are just less tolerant than they used to be. I am spending most of my time working out schedules for the field people so they can be in two places at the same time."

Barbara Wilson, retail merchandising: "All the above are my biggest problem. We receive the wrong material at the wrong time. We have calculation errors made by our sales clerks, we have customer returns that are growing higher every day. All my time is spent traveling from one store to another urging the managers to work harder on bringing customers into the stores. They keep on me to get them better merchandise and more adept clerks."

Everyone sat silently for a moment. Emory glanced at Becky, who shrugged her shoulders with a silent "I told you so."

"Well, let me ask you how you think our quality is. Do we need to do something about it?"

"Our quality isn't any worse than everyone else's."

"We are spending a fortune on inspection right now. If we pay any more attention to quality, we will go broke."

"Quality is important, but if we don't ship on time we will lose customers. We need to get into statistical quality control. That will keep everything straight."

Everyone was silent. Spellman looked at them and asked, "Do you feel that management is the big problem here?"

A chorus of "no's" swept the room. Some thought that training on quality for the supervisors might be a good idea. One suggested (in jest) that they hire Japanese workers. But overall the consensus was clear that this was just a result of the social and business level of the day. They were probably doing everything that could be done.

Harrison Ellis did have a comment. "I think all the talk about quality today is overdone. It just isn't possible to get everyone to do everything right the first time and you have to protect yourself.

"We are making progress in some areas in reducing defects, and I think there's a good attitude among the employees, particularly in production. But we have to be practical and recognize that we live in the real world and in that world errors are common."

Emory excused the group and turned to his guide. "It isn't as bad as you thought. They seem to be thinking along the right lines."

"Really. Of course it doesn't make too much difference along what lines they think; the bottom line is, what do the customers think about CC? Would you like to hear a little of that?"

"How does that happen?"

"Turn on that TV in the corner."

Spellman picked up the control device from his desk drawer and pressed it. As the set began to glow he asked, "What channel?"

"Doesn't make any difference," she said. "And you can talk to the customers, ask them anything you like."

A young mother, standing in her kitchen, hands on hips, glared into the camera.

"Are you the person who runs the company that made my refrigerator?"

"Yes, ma'am, I am," replied Emory.

"Well, you should be ashamed of yourself. I have had the service people out here three times since we bought this machine last month. And right now it is sitting there defrosting all my food because something isn't working right and they have to air express a part in. My old one worked for years. I wish I had it back."

"I'm sorry you are having trouble. Are the service people treating you right?"

"They do fairly well considering the pressure they are under. Apparently, appliances are failing all over town. But we won't have the problem much longer. We're sending it back tomorrow."

The screen faded and then brightened as another face appeared.

"I was told that the head of CC was going to interview me. Is that true?"

"Yes, sir, this is Emory Spellman. Are you a customer of ours? What product did you buy?"

"I didn't buy any product. I use your credit card. In fact, my company has thirty-four of your cards. We give them to the traveling people and it cuts down on the cash usage."

"That's a good program. We have been very satisfied with the way it developed."

"Well, I'm glad you're satisfied. Our comptroller has been jumping up and down all year because each month there are a dozen errors in the statements CC sends us. And they are all mistakes made by your people.

"Then, while we are working to get it straightened out, we get this stream of telegrams saying you are going to sue us if we don't pay up. You really should do something about that, if you want to keep the account."

Spellman agreed enthusiastically. "Let me get your name and company and I will make absolutely certain that this gets resolved."

"It's the Hartford Window Frame Company and my name is Wilson Weatherford. Good luck, you'll need it."

Spellman called the head of the credit card company and told him about the problem. He was promised an answer within the hour.

He turned back to Becky. "I am beginning to get the idea, although it seems to me that many of these problems are just part of life. What can be done about it? It is obvious that my key people don't think it is their problem."

"Well, that is one of the biggest problems," said Becky. "Quality improvement is a process, not a program, and it takes a long time for it to become a normal part of the scene. The first thing that has to happen is for all the 'thought leaders' of the company to understand quality the same way."

Spellman made notes. "Okay, we'll get everyone together and have Harrison give them a talk on quality. What's next?"

Becky smiled and shook her head. "In the first place, he is primed to overflowing with the 'conventional wisdom of quality.' He is a big part of the problem. The level of people I'm thinking about need outside education. They need to understand the Absolutes of Quality Management. They have to get in touch with reality, and so do you."

"I don't have any problem doing that—but what's the difference between what you called 'conventional wisdom' and what we need to know?"

She handed him a little card. It read, "Convention says that quality is goodness and therefore something vague; reality says that quality is conformance to requirements and therefore very specific.

"Convention says that quality is achieved through inspection and testing and checking; reality says that prevention is the only system that can be utilized.

"Convention says that the performance standard for employees should be acceptable quality levels or 'That's close enough'; reality says it must be specific, like error-free, or having zero defects.

"Convention says that quality should be measured by indexes and comparisons; reality says that we should calculate the price of nonconformance."

Spellman studied the card. "That doesn't sound too tough. I pretty much agree with all that. I think error-free probably is a little strong. That sounds unrealistic."

Becky smiled. "Most executives think that way. Let me show you the warranty record from one of your competitors. You are averaging 4.1 service calls the first year on your refrigerators; they are averaging 0.02 calls. That is a very big difference. And, before you ask, it is all 'apples and apples.' Not only that, but they are having a meeting today at which the chairperson is requesting that they get the 0.02 down to zero."

Spellman was amazed. "Has this been going on all this time?"

"The comprehension to bring about this improvement has been known for some time. The willingness of management to do it has not. They are a stubborn breed."

The telephone rang and Emory excused himself to answer it. The head of the credit card operation reported that the number of errors in the Hartford account had dropped from a high of thirty-seven a month to its current level of twelve. He also reported that the customer was satisfied with the progress. Spellman said he doubted it, thanked him, and hung up. After a moment he turned to speak to his guide, but she was gone.

The door opened and a rather sloppy individual strolled in, accompanied by a severe-looking person carrying a briefcase and dressed in a black three-piece suit.

"Are you Spellman?" asked the sloppy one.

"Who are you?" said Emory, rising to his feet. "What are you doing in my office?"

The severe one cleared his throat. "I'm afraid the office belongs to Mr. Blanton here. He just purchased the entire corporation from the bankruptcy court."

Emory stiffened and then smiled. "Okay, you two, out, out. I don't need the 'spirit of quality yet to come' to help me get the message. This company is not going to ever go bankrupt."

Obediently, the spirits disappeared.

"I am determined," said Emory Spellman to himself, "that we are going to solve this quality thing."

He looked at the card Becky had given him and turned it over. On the back it said, "The quality policy of this company is that we will deliver defect-free products and services to our customers, both internal and external, on time."

"I'm beginning to realize," said Emory to himself, "that I have had a bad case of tunnel vision. I limited myself on the idea of quality. Quality is not just making things like requirements and being steadfast in insisting on doing things right. It isn't a functional thing at all. Actually it has to do with the way the company is run overall. It takes the combined actions of every person in the company to cause the company to operate properly.

"What I want is not just zero defects in the products, and happy customers, and profitability, and all that stuff. We need it, goodness knows. But that is only a result.

"What we are going to make happen around here is a hassle-free company. We are going to learn how to do all the things we do in an orderly, energetic, and accurate fashion.

"Hassle-free. That's us."

And he set off to do just that.

# 5
# Determination

Companies don't do well with quality because they are just not determined enough. I realize this statement sounds as profound as, "It is better to be rich and healthy than poor and sick." However, it is true.

Earlier in the book we discussed the profile of a company that always has problems with quality. Nothing changes until a company breaks that lock.

Quality improvement also has a profile. The companies that don't get much improvement, even though they appear to be determined, have common characteristics:

1. *The effort is called a program rather than a process.* This reflects the idea management holds in its secret heart—that this quality business is one of finding the proper set of techniques to apply to the proper people. A "program" lets people know that if they wait and go through the motions, it will soon be replaced by something else. Governments call everything programs. A "process" is never finished and requires constant attention.

2. *All effort is aimed at the lower level of the organization.* It is easy to identify this situation. Just try to find something that senior management has to do differently: all the schooling is for someone else. The Productivity (with a capital P) efforts all are for low levels. Quality circles never begin in the boardroom. Statistical quality control is not applied in white-collar areas.

3. *The quality control people are cynical.* "Zero Defects is Eastern mechanical thinking." "We have to satisfy the customer's perception of quality." "It just isn't possible for people to do things right the first time." "The economics of quality require errors; you have to consider the trade-offs." All the above are part of the conventional and cynical wisdom that has kept nonconformance an integral part of business. Fortunately, the swing in the quality control profession is toward the reality contained in the Absolutes of Quality Management.

4. *Training material is created by the training function.* The concepts of quality improvement and the actions required to cause it are very subtle and require a comprehension that comes from experience. However, they sound so simple that people get right into training without realizing that they are actually teaching the ideas that caused the problem in the first place.

5. *Management is impatient for results.* As soon as it learns about the cost of quality, management notifies everyone that it expects an immediate reduction. This results in a lot of short-range actions, like shrinking the quality department.

We have learned that there is a 25 percent reduction in nonconformance costs in the first year if the process is properly applied. There is usually an increase if it is not properly understood.

Impatience also leads to centralization of the program. This means that the individual managers lay back and wait for the word to come down. That brings everything to a slowdown since it increases the hassle.

These characteristics, and a few more, show up in the poorly run quality improvement process. They occur because the entire event has not been thought out and taken seriously enough.

In establishing its determination, management has decided, perhaps without knowing that it has done so, that everyone else needs to do something different. It is not until management decides to clean up its act that real determination sets in.

Football fans recall that a famous coach became disturbed because the opposing team's fullback was making a great deal of yardage over his middle guard. Turning to the bench he called out, "Smith, go in there and stop that guy!"

Pulling on his helmet, Smith prepared to dash onto the field. "I'll try, coach," he said.

"Sit down, Smith," said the coach, "the guard I have in there now is trying."

This is the type of story that motivation sessions are built around. And there is a great deal of truth in it. "Trying" is not enough. "We want someone who will do the job." Dedication, determination, drive are necessary if something is to happen.

But this is not a discussion of motivation. I never have felt that you could "motivate" anyone for more than a few days. This is all about what makes quality happen or not happen in a company or organization.

I once was a member of the quality department in a company that had terrible quality. The quality department passed every audit the customer could devise and its manuals and procedures were held up as examples of how these should be written. It was indeed a good operation full of good people. It took its responsibilities and skills very seriously.

However, its main skill was finding nonconformances before they could be delivered. We had rooms full of them. As fast as we found and impounded the tainted material, management figured out new ways of accepting it. The company wasn't trying to do wrong; it just felt some of the requirements were not that critical.

As a result, the product didn't work very well, the customer became dramatically unhappy, and we got a new general manager. Of course, we had gone through four quality directors first. That was the past GM's way of dealing with the problem.

The new GM made it quite clear that we were going to produce in strict accordance with design and that when something met all aspects of said requirements, we would deliver it. In the meantime, we would identify every problem that needed fixing and we would fix it.

"Fix it" included figuring out how to keep it from happening again.

Two senior people thought the GM wasn't serious and led their operations on a wavering path. The GM fired them. From then on, everyone worked his way. I had a problem list of fifty-six items when we began, and it took six months to work them off. Along the way, another twenty-four were added and dealt with.

At the end of six months we were back on schedule, the material review cages were shrunk to closet size, and the customer was running test after test trying to find something wrong. But nothing was wrong. It was hard to believe, but the product actually worked; it was reliable, within the budget, and on time.

Here was a case where the exact same organization produced two opposite results. The only difference, outside the two fired hardheads, was the leadership.

The GM made it crystal clear, day after day, that he was determined to produce quality in the truest sense of the word. He stalked those who did sloppy housekeeping; he invited the hourly paid people and their families to evening dinners in the plant; he stopped by the material review crib and took defective parts up to the chief engineer's office; he took the union president to lunch; he chased the directors out of their offices and into the real world; he made a real pain in the tail out of himself. No one was allowed to forget, and the new quality director gleefully helped everyone learn how to prevent nonconformance.

American management has almost worn a rut between here and Japan searching for the secret of Japanese quality. "Why does their stuff work when ours has so many problems?" "Why can their U.S. auto dealers spend their time selling instead of fixing?" "Why do all their administrative actions get done correctly and we have so many problems?"

So Americans go to Japan, and they return to install some magic tool that will fix the quality problem. They come back with reinforcement for the idea that the problem is the worker and that what is needed is some way to get American workers to act like Japanese workers.

Unfortunately, little happens in the way of permanent improvement. Not because there is anything wrong with the techniques utilized; they are all worthwhile. But little improvement will happen until the real problem is dealt with. The real problem is that management doesn't take the product and service requirements seriously.

If you look behind the Japanese success, or the success of the many American companies that have quality products, you will find one secret: they take the requirements seriously. They create them with care; they meet them with care. They don't use "off specs," "that's close enough," "waivers," "mail it next month," and such.

When management insists on conformance to requirements and provides the participation necessary for prevention to happen, it becomes a different world. All the quality awareness procedures ever developed can't make that happen. It won't occur until the employees of the company have a common language of quality and understand what management wants.

It is easy to tell after fifteen minutes inside a company how closely its administrative activities resemble the procedures and, if it is a hardware company, what its quality level is. I never miss it by very much. In some situations all this reveals itself after a few minutes in the chief's office. The operation looks exactly like the management.

It is not enough to look determined and act determined. The subject we are being determined about has to be clear in the minds of all involved. How that clearness gets there is covered in the chapter on education. But nothing flourishes unless it is clear in the minds of the "thought leaders."

We have learned that these people cannot be taught inside a company. They respond only to credibility that cannot usually be accepted inside a peer group. They need to go off-site to be subjected to an immersion that will help them change their minds.

We have also learned that quality improvement efforts will only work with companies that come to it voluntarily. You can't go out and drag them in. This strategy is the result of having learned that management has to make its "born again" quality commitment on its own. The senior executives go through the education system and learn their role in the whole process. They do this and they learn.

However, when the executives, managers, and quality improvement people enter the education system, they normally want some reassurance that management is really committed. It isn't that they don't believe in its seriousness; it is that they are not certain it will last. As soon as the current problems are relieved they think the "old man" might get off on another kick.

The credibility of the commitment is the biggest single problem for management; it has to be reinforced all the time. Management has to continually show it is in it for the long haul—forever. It is not enough to say the right-sounding words; everyone does that. The actions and the lifestyle have to be visible. (This is what the Japanese senior managers excel at doing. Never will one hear a discouraging word about quality from those folks.)

It is impossible to hide or fake clear understanding of a subject. Real comprehension has to exist if the determination is to be obvious.

That brings us to the Absolutes of Quality Management.

As a professional quality manager in a world where no one is against quality and yet very few have it, I tried for years to help comprehension along. But nothing worked dramatically because everyone thought they understood it all. It is difficult to reach the mind of someone who is enthusiastically agreeing with you. It is hard to corral the coworker who feels that the two of you are the only people who are "good."

Finally, the Absolutes came together as the four basic concepts of the quality improvement process. There is a fifth: "There is no such thing as a 'quality problem.'" But that is for the professional quality people, and very few of them understand that its purpose is to pinpoint problems more tightly than, "The quality department is a mess."

The Absolutes answer the questions:

1. What is quality?
2. What system is needed to cause quality?
3. What performance standard should be used?
4. What measurement system is required?

# 6

# The First Absolute: The Definition of Quality Is Conformance to Requirements

## IMPROVE QUALITY AND ELIMINATE HASSLE AT THE SAME TIME

Quality improvement is built on getting everyone to do it right the first time (DIRFT). Causing that to happen is what this book is about. But the key to DIRFT is getting requirements clearly understood and then not putting things in people's way.

Management really has three basic tasks to perform: (1) establish the requirements that employees are to meet, (2) supply the wherewithal that the employees need in order to meet those requirements, and (3) spend all its time encouraging and helping the employees to meet those requirements.

When it is clear that management policy is to DIRFT, then everyone will DIRFT. They will take requirements as seriously as the management takes requirements.

Hassle comes about because of vacillation in management's dedication to the policies and processes. When no one can count on anything, then no one plans on DIRFTing. The "it" in doing it right the first time is the requirement. No "it" means no chance to DIRFT.

One of the first business jobs I ever had was as a technical editor. The company was writing descriptions of material owned by the government. My job was to write the descriptions in such a way that anyone with a low technical comprehension (such as I) could understand it.

I would do the research and then the writing. My material would go over to the evaluation group. They would review it and then come over and yell at me. The work was never right.

After a few months of this I finally made up a list of the things I thought the technical description was supposed to cover. The groups reviewed this list, added a few things, and changed some. We shook hands on the list and from then on we lived happily in our work.

We had agreed on the requirements and they were ours.

Now, this may not sound like a development along the lines of the invention of the telephone or walking on the moon. But it took a situation full of hostility and hassle and turned it into a pleasant, productive work situation.

We agreed on the requirements. Before that it was everyone for himself or herself, and the higher in the organization you were the more your opinion counted. No two reports were the same. The energy of the organization was dedicated to resolving the same situations over and over. The instinctive feeling was that things could not be defined in order to be accomplished routinely.

Such activities cost service companies a conservative 40 percent of their operating costs. Think about that as we go on to why it happens and what to do about it.

The cause is not the techniques of processes or procedures; it is not a lack of knowledge of how to do things; it is not a lack of desire on the part of the workers to have orderly lives. The cause is management's definition of what quality is: goodness. Nobody knows what that means except the speaker.

Quality has to be defined as *conformance to requirements*. This definition places the organization in the position of operating to something other than opinion and experience. It means that the best brains and most useful knowledge will be invested in establishing the requirements in the first place. They will not be used in determining what can be done to smooth over the rough places.

The determined executive has to have a brain transplant where quality is concerned. When someone comes rushing up with something that has been turned down for having a small imperfection, the determined executive has to snarl something like, "Why would we want to send something to our customers that isn't what they ordered?"

The determined executive has no recourse except to make the same point over and over until everyone believes. The first time a deviation is agreed upon, everyone will know about it before the ink is dry. "Oh," people will say, "there are some things that don't have to be right."

A group executive in a computer company realized one day that a new product was scheduled to be released from development into production. However, a recent status meeting had shown that the new product had not finished the required testing. Release of it would mean that once again development would be completed during the manufacturing process. In turn, manufacturing would not be able to finish its tasks completely, and the field service people would be called upon to make the product work in the customer's office. The bugs would be out in a couple of years.

This arrangement was taken for granted, since it had happened for years. The group executive knew how much this system cost and how often the customers had been disappointed by the performance of new products.

So the decision was made not to meet the final delivery date, which was scheduled for several months down the line. Development would not be certified as complete when it was not complete. The result would require juggling by marketing, returning some customer deposits (with interest), and losing some face in the industry.

But they did it anyway. The effect wasn't as bad as expected. The product was only a little late, since production development didn't take as long when they had the opportunity to work with clean paper. The field service requirement for repair practically disappeared. The product's reliability was the highest to date in a new line.

After that, new products routinely completed development on time. No additional cost was necessary, and field reliability was even higher. The procedure became routine once everyone realized that management was going to stay serious about it.

It is not necessary to shut the company down to prove you are a determined executive. It is possible to say, "I will give you six weeks to clean up this situation. In the meantime, I will let the current 'doing it over' process continue. However, when the six weeks are up, if it isn't corrected forever, then you can just eat the product."

## QUALITY GOLF

Greg Howland braked his car to a stop in the club parking lot, pressed the "open trunk" button in the glove compartment, opened the door, and ran around the car. He quickly stepped out of his loafers, struggled into his golf shoes, and reached to pick up his golf bag.

Tom Morley drove up to the car in a cart.

"You're just barely in time, as usual, old buddy," said Tom. "Put your bag on the cart and let's go. We're on the tee now. Today is a full handicap tournament.

As Greg finished dressing, Tom drove over to the number one tee where their two friends were waiting for their Saturday morning game. Greg endured the good-natured teasing and waited for his turn to hit.

Reminding himself to remain calm, Greg swung uncalmly at the ball and drove it into the driving range on the left side of the course, out of bounds.

"Hit a Mulligan," said Tom. "We'll take pity on you since you were late again."

Greg drove down the left side of the par five hole and the group went forward. Reaching his ball, Greg discovered a clutch of practice balls sitting beside it.

"This will give me a chance to warm up," he said and proceeded to stroke five balls over the fence onto the range. Then, addressing his ball, he noticed a large weed immediately behind it.

"If I hit that weed, it will kill the shot," he said as he yanked the offending plant from its roots. Then he was able to press down the ground behind the ball.

The shot went smoothly down the middle of the fairway, landing near the second shot of his opponent.

"What club are you going to hit, George?" asked Greg. "It looks like about 150 yards to me."

"That seems about right. I'll use a six iron," said George.

After George's shot fell short, Greg hit a longer club over the green into a sand trap. He drove to the trap muttering to himself about the difficulties of figuring distances on this hole.

Flicking a pebble from in front of his ball, Greg blasted from the sand and fell nicely on the green, lying four. When his turn to putt came, his partner held the flag stick for him.

Thinking the ball was going past the hole on the right, Greg's partner released the pole and walked back to his ball. However, the ball broke sharply as it reached the hole area and fell into the hole, striking the stick and bouncing out to hit on the edge.

"Tough luck," said his opponent and knocked the ball back to him.

What was Greg's score?

Greg will tell you he had a six on that hole and would have had a five if his partner had not been so lazy. However, golf is a game of rules, and if you follow the rules, Greg had a few more than six.

|  | Rule | Accumulated Total |
|---|---|---|
| a. Drove ball out of bounds |  |  |
| Penalty: stroke and distance | 29 | 2 |
| b. Agreed to waive the rules |  |  |
| Penalty: disqualification | 4 | 2 |
| c. Drove ball |  | 3 |
| d. Practiced on the course |  |  |
| Penalty: two strokes | 8 | 5 |
| e. Improved lie |  |  |
| Penalty: two strokes | 17 | 7 |
| f. Hit ball to mid fairway |  | 8 |
| g. Asked advice and assistance |  |  |
| Penalty: two strokes | 9 | 10 |
| h. Hit into hazard (sand trap) |  | 11 |
| i. Moved loose impediment in hazard |  |  |
| Penalty: two strokes | 18 | 13 |
| j. Hit onto putting surface |  | 14 |
| k. Putted to the hole |  | 15 |
| l. Hit the flag stick |  |  |
| Penalty: two strokes | 34 | 17 |
| m. Failed to hole out |  |  |
| Penalty: disqualification | Definition #3 | 18 (with "gimmie") |

By the published and agreed-upon rules of golf—those we have all agreed to play to every day—Greg had eighteen strokes instead of six and was disqualified twice.

This means that he was playing to a set of requirements different from those established by the rules of the game. As a result, his scores were not golf scores.

If we do not take the requirements seriously, then we will not perform the task well. If Greg knew he had to live with every result instead of being able to adjust the situation to smooth out his misfortune, he would play differently.

The professionals do. They practice hitting the difficult shots as well as the easier ones. They warm up before going out on the course. They work on their attitude so that when adversity strikes (like a ball in the lake) they can handle the problem without making it worse.

The difference between the professional golfer and the amateur is not just the swing. Much of it is an understanding of the game, beginning with the rules. It helps their discipline. If all the players make up their own rules as they go along, is it any wonder that their games don't improve over the years? Taking requirements seriously is the first act in improvement.

The first Absolute of Quality Management is:

QUALITY HAS TO BE DEFINED AS CONFORMANCE TO REQUIREMENTS, NOT AS GOODNESS.

Managers tend to get very worried when the subject of setting requirements comes up. They immediately visualize thousands of little "Do this and do thats."

Yet if you think about requirements, they are really only answers to questions. It all begins when you tell your people that you want them to "Do it right the first time" and they want to know what the "it" is.

What time do I come to work?

*Eight-thirty in the morning.*

What time do I go home?

*Five o'clock in the evening.*

Where do I sit?

*Here.*

What is my job?

*Accounts payable.*

I know how to do that from the place I used to work. Do you want me to do it that way here?

*No. Here is the AP manual. Please look it over and see if you think it will cover everything. We want to pay all our bills on time in order to get the discount and keep our credit rating.*

What input will I receive?

*Accounting will send you all the bills as they are approved for payment.*

How do I know how much money is available for payment?
*Here is the computer operating manual for this function. Ms. Sneldly will come here tomorrow morning at 8:45 and show you how to operate the terminal and printer. She will stay as long as you want her to.*

Do I pay every day?
*We like to pay each Wednesday, but you can make up special checks if you have a good reason.*

What is a good reason?
*Here is a list of good reasons. One is the chance that we would miss a discount by waiting.*

How do I address the checks?
*Every statement has a "Make check payable to" line on it.*

And so forth until all the questions have been answered.

In another area you might hear someone asking, "How deep should we make this hole, how wide, what tolerances should we use, what type of finish, what kind of material . . . ?" The answers to those questions produce requirements.

Requirements, like measurements, are communications.

# 7

# The Second Absolute: The System of Quality Is Prevention

The most visible of the expenses of conventional quality practice lie in the area of appraisal. Manufacturing companies label people inspectors, testers, and such. Service companies have the same activities under different names. The big difference is that in manufacturing these people are identified, trained, and led. They become a potent force for discovering problems and helping cause corrective action. They have a companywide impact.

In service companies, those actions are disjointed and usually not pulled together. It is very difficult to identify companywide problems. It is harder to cause corrective action or even to be listened to. There is no difficulty in arranging these duties; it is just that the service culture is not used to doing it.

Appraisal, whether it is called checking, inspection, testing, or some other name, is always done after the fact. If these actions are used for acceptance, then the result is sorting the good from the bad. Each action produces a little pile of material or paper that has to be further evaluated.

For a clear dramatization of this, visit a machine shop or some similar hardware operation. Each batch of work usually moves around in a "tote" box. Each box is accompanied by a folder that contains the drawings and specifications applicable to the material in the box.

On the outside of the folder is a listing of the routes that the box is to take. Each step of the process is listed: shear, drill, inspect, deburr, shape, inspect, and so forth. All this is laid out by the industrial engineer. The routing sheet has a place for writing down how many pieces survive each operation. If the box began its journey with 100 pieces, one might think it was expected to end its journey with 100 pieces. Not so. After each step, the operator involved marks down how many are left.

When the defect levels become known, the stock crib learns how many items have to be placed in the box in order to produce the desired amount, perhaps 110 to get 100.

This type of thinking is one of the major reasons many of America's basic industries are in deep trouble. Multiply each little box of 100 by tens of thousands, and it is not difficult to come up with an incredible amount of waste.

Service people do the same thing by overloading the amount of people necessary to do jobs. Each bump in the road soon has an attendant. Data systems have backup systems. New forms have newer forms. Review meetings have other meetings.

Appraisal is an expensive and unreliable way of getting quality. Checking and sorting and evaluating only sift what is done. What has to happen is *prevention*. The error that does not exist cannot be missed.

Prevention is something we know how to do if we understand our process. If a salesperson is driving from a strange airport into a strange town, it is best to ask for directions prior to heading out on the highway.

If a painter wants to match a color, it is best to take a sample to the paint mixing store.

If a restaurant owner wants fresh eggs every morning, it is necessary to locate someone who sells fresh eggs and have them delivered on time.

If a buyer in a retail store wants to receive the correct sizes to meet the customer requirements, there is only one way to do it. The producer must receive an order with the selection of sizes clearly laid out in a communication that can be mutually understood.

It is hard not to agree with these commonsense actions. It is much too late to check sizes when the boxes arrive, to open each egg and look inside, to run back and forth trying to remember colors, or to peer at a road map as an eighteen-wheeler rides the rear bumper.

How come such thinking is not part of the normal operating system? Why do we lurch from post to post and back again instead of working out how to do it right the first time, every time?

Prevention is one of those things that business people just don't talk about. The subject of "Let's do this one right," or "Think this one through first," comes up regularly. But no one takes that seriously.

When I first became a department manager, I worked hard on getting my budget laid out just right. It was tight, but I felt we could make it. During the year we stayed right on track even though my people had to do without many times. At midyear, we cut back on travel, which slowed down some projects. But I was looking forward to hearing the boss tell me what a good job I had done when the year ended.

My friend, who ran test, paid absolutely no attention to his budget. He did whatever he wanted to do and wrote great justifications. At year-end he was 20 percent over budget. I was right on the money.

When the budgets for the new year appeared, my department took a 10 percent cut and his had a 20 percent increase. That was the last time I ever took such a budget seriously. The credibility of management was damaged forever. I overran the budget by 20 percent the second year and no one said a word about it. It is by such experiences that we learn that management is really not interested in prevention.

Yet we who are now in management know that the old days are gone. No one can afford to blunder along anymore.

The concept of prevention is based on understanding the process that needs the preventive action. Whether you are making printed circuit boards or preparing insurance policies, the concept is the same. "Making big ones out of little ones," my father used to say.

The secret of prevention is to look at the process and identify opportunities for error. These can be controlled. Each product or service contains many components, each of which has to be dealt with to eliminate the causes of problems.

An insurance policy, for instance, involves a series of opportunities for error, say there are twenty-five of them. The salesperson needs information that will show the prospective client the benefits and fees involved. The training people need accurate data to pass along. The development and preparation of that information is a major task in itself. The actuarial and marketing input also is very large, and any errors or omissions will be expensive for someone. The salesperson must transmit the client information to the place where the transaction begins. Here one little transposed number will set a computer off on the wrong track. The forms and the systems involved have to be designed in a way that eliminates the opportunity for casual errors.

In one case a casualty insurance company's chairperson was being besieged by several major independent agents. They were upset with the clerical errors that were being made. Letters were mailed to the wrong places, client names were misspelled, engine numbers were transposed. There was always something.

The company first decided to set up a room with 200 people in it to check all the packages before they went out. Considering that this method would cost about $5 million and guarantee very little added confidence, it wasn't hard to suggest looking for an alternate solution.

An alternative was developed after spending some time with the independent agents. There were about 150 involved, and they represented a large percentage of the firm's business. It was determined that most of the policies were fairly standard, such as automobile liability. It was these routine policies that always had problems. When someone wanted to insure an elephant, or do something special, it was handled with little trouble.

The company set up computer terminals in each agent's office. This let each agent's people type in the information on the policy order blank and transmit it to the insurance company's computer. After a moment or two the computer would transmit back, the policy would automatically be typed out, and everyone was happy. Any mistakes in spelling or numbers belonged to the agent. That took care of the chairperson's problem.

This solution really delighted marketing, because it caused the agents' offices to place more business with the company. The firm is now placing the terminals everywhere possible.

Here massive opportunities for error were eliminated, and that is what prevention is all about. Those terminals cost the company a fraction of what the checking function would have required. The elimination of mailing expenses alone paid for much of the installation.

In manufacturing processes, particularly assembly or high-production operations, there is a technique that helps prevention. This is SQC, or statistical quality control. Under this method each variable in the process is identified and then measured as the process continues. When a variable begins to move out of control, it is adjusted back in. If all the variables are inside their lines, then the end result should be just what was planned.

SQC is made out to be very complicated and difficult to do, but there really isn't that much to it. It is a very effective, easy-to-understand tool. The people who make up the charts for the variables and teach the measurement techniques have to be skilled, but everyone else only has to learn to understand a few things.

The charts are made up with an upper and lower limit representing the "tolerance" of the process. Each measurement is recorded by a dot on the chart. If a dot is within the lines, keep running. If the dots are heading outside the lines, do something. If a dot is over the line, shut down. That is all there is to it.

However, very few managers can accept this. They just can't stand to interfere with the process. For this reason, SQC has never been accepted in the U.S. as a normal part of operating; it has always been considered a special step. Right now many companies are using SQC, as they should. But for many it interferes with the business of being resourceful.

Executives do not realize the effect their personal actions have on the processes of their company, whether it is a service or manufacturing firm. (The only difference between the two is that the waste in service companies goes out in baskets and in manufacturing companies in barrels.) Remember that the majority of the people in hardware companies never touch the product. They are all in administrative and other paperwork functions. The best laid plans for prevention can be undone by a careless executive policy. I can think of one clear case.

I like to visit operations, whether they are the back room of a stock exchange company or a foundry. I like to see people involved in doing things and to have the opportunity to learn from them. On trips like that, I pick up a lot of information that later distills itself into a new understanding.

One day I went to see a company that was becoming a new client. As part of my orientation I was taken to see a new process whereby a new component was being produced. The dust-free room had been newly constructed and the complex process had been developed at a cost of several million dollars. It was very impressive.

My host fitted me out in white coveralls, cap, gloves, and shoe covers. We toured the line chatting with the people working there. I asked several folks what their biggest problem was and received a varying supply of answers. The morale obviously was good, and the people were proud of their work.

"This process has twenty-one different stations," my host said. "Each one is monitored electronically and the status is reported continually. For instance, this bath right here has to be a certain temperature. The digital readout shows what it is whenever this button is pressed. It will even give you a printout."

Real-time data always interests me since that is one way to control a process closely. I considered the process to be an achievement.

At the end of the line my host took me to a desk where the process control technicians were headquartered. He showed me the report slips that they completed after their rounds.

"The process stations are monitored regularly, and whenever any requirement is out of the control limit, a card is written. Then the process engineer looks into the situation and determines whether the control point has to be corrected, and if it does, when.

"We find that we can do most of our fixing and maintenance on the third shift."

I examined the cards and commented that each had been noted, "Continue process" and signed L. S. Jones.

"Jones is the process engineer," he said.

"So the process is continued and then adjusted if necessary at the end of the shift?" I asked.

"Right. That way the maintenance work doesn't interfere with output. There is a big demand for this device and we are behind schedule."

That night I had dinner with the chairman, president, engineering director, manufacturing director, quality manager, and personnel manager. They took me to a very nice restaurant and we had a good discussion. The chairman and I had not met before, and he was very interested in my thoughts on quality improvement in the industry. He felt that a great many of the problems they were having were due to the price advantage the Asian communities had. He didn't feel that the reliability differences were real.

I commented that my electronic equipment clients reported they could take semiconductors from some suppliers right from the box and use them. Others had to be tested extensively. The chairman's company had a bad reputation for quality and he knew that.

As coffee was served the chairman asked me to make whatever comments I liked. He was particularly interested in my impressions of their new product operation.

I realized that this group was concerned about quality improvement. They really felt the need. However, I also could tell that they felt the problem was primarily one of worker behavior, product design, and competition that received cheap money from its government. They needed to realize that they as individuals, and as the thought leaders of the company, were causing most of the problems.

"I am very pleased that we were all able to get together for dinner," I said. "It is very helpful for me to be able to get to know all of you and to have seen some of your processes.

"I must say, however, that I am surprised you didn't invite L. S. Jones to dinner with us."

The chairman looked at the president. "Who is L. S. Jones?"

Before he could answer, I jumped in. "He's the guy who runs the company," I answered.

"I run the company," said the chairman. "All of us here do, plus a couple of other people."

"Jones is a process control engineer," said the quality manager. "He works on the new product line. A very good person."

"Why do you say he runs the company?" asked manufacturing.

"Well, he decides how good your products are going to be every day. His primary objective is to produce components, so if something goes wrong that is not catastrophic, he keeps things running. And he does this because that is what you, management, have told him you want."

"All new processes have start-up problems," said the chairman. "If we waited until everything was perfect, we would never produce anything. We have to deal with yields."

"Your company spent millions of dollars to create a controlled process to make these components. Then each time one of those controls tells you there is a problem, it is overridden. I would be very surprised if the yield from that line were 35 percent."

"It is 16 percent," said quality.

"But we have to build these yields up gradually," said engineering. "We don't always know everything about it when we begin."

"And you never will understand this process until you work it a little. I seriously doubt that there ever have been two days' runs that were close together."

"What are you asking us to do?" said the chairman.

"First, remember that all the problems are caused by management. Second, run the line according to the requirements that have already been established. If something goes out, stop and fix it. It will all run smoothly soon."

There was silence.

"I presume we will all understand this in depth once we have been to the Quality College?" said personnel.

"That would be a big help," I replied.

"Why don't we all go meet Mr. Jones in the morning?" said the chairman. "I need to know more about all this."

The next day we all put on white outfits and toured the line together. The chairman had the opportunity to meet L. S. Jones, who confirmed what I had surmised. After the tour we sat in a conference room for a few moments.

"I think we ought to give this a try," said the chairman. "It seems so obvious, it makes sense. I suggest that we only run the line when all the process control points are in control. We also should learn how to keep them in control so we can run all the time."

They did just that and in a month were running a yield of over 85 percent. As they learned more about the process, they raised that and began to improve the design.

The difference between 16 percent and 85 percent required no new people, machines, money, or anything except determination. The schedule problems disappeared and new products were introduced in the same line due to increased capacity availability.

The second Absolute of Quality Management is:

THE SYSTEM FOR CAUSING QUALITY IS PREVENTION, NOT APPRAISAL.

# 8
## The Third Absolute: The Performance Standard Is Zero Defects

Setting requirements is a process that is readily understood. The need for meeting those requirements each and every time is not so readily understood.

A company is an organism with millions of little seemingly insignificant actions that make it all happen. Each and every little action has to be done just as planned in order to make everything come out right.

The performance standard is the device for making the company happen by helping individuals to recognize the importance of each one of these millions of actions. When a company encourages people not to do everything right, they cause some of those actions not to happen. No one knows exactly what will or won't occur.

The *New York Times Magazine* of August 29, 1982, contained an article by Jeremy Cherfas and John Gribbin about updating human ancestry. The authors related the results of their study of history through examination of the molecules of the different species. Instead of examining the bones and artifacts of past civilizations, they were looking at the molecules of the bodies of living organisms. They determined among other things that a human, a gorilla, and a chimpanzee are siblings from a biochemical standpoint. In fact, their DNA is so similar that there is only 1 percent of difference.

DNA (deoxyribonucleic acid) is the component of every body cell that contains all the instructions needed to reproduce that complete body. If we knew how to clone, we could take the DNA from cells and set up an assembly line to produce the same person over and over. The eyelashes would be the same, the fingerprints, the heart, and everything else.

However, if we were to do all this with a performance standard that permitted a certain amount of error, then we might find ourselves with part gorilla and part what we began with. We certainly wouldn't have the identical person each time.

A company with millions of individual actions (think of how many different actions you personally take each day) cannot afford to have a percent or two go astray. Less-than-complete compliance with the requirements of a performance standard could cause that to happen.

Companies try all kinds of ways to help their people not meet the requirements. For instance:

- *Shipped-product quality level (SPQL).* This means that a certain number of errors are planned. Refrigerators have perhaps three or four, computers eight or more, TV sets three or more. The purpose of having a shipped-product quality level is to let management determine how many field service people will be needed.

- *Acceptable quality level (AQL).* Usually established for suppliers, this number (1 percent, 2.5 percent, or some such figure) is supposed to establish the acceptance plan for inspection or test people. However, it really represents the number of nonconforming items that can come in an acceptable lot.

These and other variations on the theme convince people that the determination to get things done right just isn't there. I have listened for years as otherwise reasonable people explained and explained how Zero Defects was an impossible goal. Yet, in their own companies, there were areas that routinely had no defects.

Check the payroll department and see how often an error pops up. Whenever a problem comes about in someone's pay, it is usually because the individual, the supervisor, or the personnel department did something.

Payroll doesn't make mistakes.

Is that because they are such dedicated souls? Certainly they are, but the importance of the work does not necessarily raise performance standards. If that were so, one would think that people working on space exploration equipment never err. However, you can get used to anything, and bad performance standards occur eventually.

The reason payroll does so well is that people just won't put up with errors there. They take it very personally when something is wrong with their paycheck. Not because they think the company is going to cheat them—they know it will all get straightened out eventually. They get upset because they feel that the company doesn't care about them if it can't even get their pay right.

Conventional wisdom says that error is inevitable. As long as the performance standard requires it, then this self-fulfilling prophecy will come true.

Visualize 100 discs mounted side by side on a bar so they can spin when touched. They would look like a round loaf of bread in thin slices, or a stack of 45-rpm records on a holder. Then visualize a wire across the front, and imagine that each of the wheels has a red mark on it that covers 1 percent of its circumference. That means that each wheel is 99 percent reliable, or 1 percent defective, depending on how you look at things. The 100 discs represent 100 components in a hardware system, or 100 steps in a paperwork process, or 100 people in a band, or whatever you desire.

When the discs are spun the system is operating. If a little red mark stops on the wire, then there is a failure. What is the probability of success with 100 steps, each of which is 99 percent reliable? Multiply 99 percent by itself 100 times. That produces a result of 36.4 percent. The success rate is going to be very low.

The purpose of quality improvement is to make those little red dots smaller and smaller until they disappear. Then you don't have to do everything over twice.

Only someone who has had the job of causing quality in an organization can realize the importance of a specific performance standard. Approaching the situation from a purely intellectual or academic posture makes it difficult to understand that employees of all levels perform to the standards of the leaders—not to the standards of a procedure or process.

In 1961, I created the concept of Zero Defects. It said we had to lay out a clear statement of what we wanted people to do. We didn't want grade levels like in schools, and we didn't want "quality levels" like in statistics.

What we wanted was to do the job right the first time. To get everyone to understand that we were serious about it required constant communication. Over the years we have learned how to do that better. Unfortunately, Zero Defects was picked up by industry as a "motivation" program.

I kept explaining that it was a management standard that told people what was expected from them, period. However, the thought leaders of the quality profession attacked this idea as being impractical. As a result, it was downgraded and ignored in most of the U.S.

The Japanese thought it was great and have been using it all these years to explain what management wants people to do. The U.S. could have been working on learning how to do things right during that time period instead of searching for that elusive "economics of quality."

Companies have elaborate reporting systems to show that they are improving. They have advertising programs that show their people working hard on quality. The only thing they don't have is error-free products.


## THE QUALITY FLYPAPER COMPANY

Quality Flypaper Corporation is well known for its flypaper all over the world. Lately it has ventured into associated consumer products such as roach paper, ant paper, and even mouse paper. Harold During is vice president, variance.

"Sticky paper is all a matter of the insect, or rodent in the case of our new products, versus the stickiness of the compound. We spend a great deal of money and time testing the sticky compound we make. My department is particularly involved in deciding where variable products are sent."

I was interested in the techniques used by the variable determination function of During's department. However, I first wanted to learn why it was necessary to conduct such determinations.

"We have found," said Hal, "that not every customer requires the same level of stickiness. This fits in well with our process, which doesn't necessarily produce the same stickiness with each run."

I asked if that meant that it wasn't always possible to tell just what was going to come out of the manufacturing process.

Hal pointed out that sticky making was not an exact science

when it came to their products. "The people who make glue or tape for offices or that sort of thing can afford to get it all the same every time. We just can't. So what we do is adapt the result to the market."

In response to my puzzled expression he reached into his pocket and removed a small gold key attached to a chain. He pulled out the bottom drawer of his desk and, glancing furtively about, opened a little chamber inside the drawer using the gold key. He remuved a leather-bound notebook and opened it quickly to one well-thumbed page. He turned the notebook and pushed the page quickly in front of my eyes. I caught a glimpse of tables of numbers as the book was quickly thrust back into the drawer and locked securely away.

"Obviously, I can't share that data with you, but I did want you to see that it existed. What I showed you is one of the most closely held secrets of the Quality Flypaper Corporation. It is what lets us be so profitable."

"But you only make 1 percent after tax," I couldn't help commenting.

He ignored that comment. "What you have just seen are the results of our worldwide study of the pulling power of a fly's legs. You see, the layperson thinks that all flies are the same. Well, they just aren't. Their capability to pull themselves free of our flypaper varies from country to country and even within countries.

"The Michigan fly in midsummer, for instance, is the Mack truck of flies. But late in the fall it is a real pussycat. However, at that time the flies of Rio de Janeiro could pull you out of bed. This is not an easy business."

I still didn't understand.

Hal leaned forward impatiently. "We can test the flypaper, or other paper that we produce, and tell immediately where it can and can't be used. If it is a little weak, we send it where the flies are weak. If it is up to specification, we can send it everywhere.

"Our lab people can tell within six hours of the completion of a production run just where the product can be most practically used. This is a tool that no competitor has."

I was a little confused. I asked whether this meant that the lab people could tell exactly what it was the company had made but couldn't tell what was being made while it was in process. Did they merely sort out the results?

"We never have asked them to look into controlling the process. We already know that can't be done. No, we concentrate on where the product we make can be used. Our new mouse paper is only effective in some of the areas of West Africa where they have had a drought for several years. The mice are pretty weak there. We haven't had much success in finding other places, but our exploration teams are out all the time."

I asked how many teams there were. Hal sat back expansively as he counted out loud. "We have nine teams out at this moment doing studies on insect pulling power and four teams working on rodents. Roaches never change much so there isn't a great deal of market for us there. We are slowly pulling out of that, as soon as we get all our warranty claims settled. Each team costs us about $500,000 a year for its work. We consider that money well spent."

I wondered whether that amount of money would be of value spent on controlling the process so they could make flypaper that would work anywhere in the world at any time. Hal didn't seem much impressed by that thought and explained to me patiently, "Quality costs you too much if you are going to try to make things better than they have to be made."

He leaned forward confidentially. "The flypaper business is different from those you deal with," he said. "Here we have to satisfy different markets. People want different things in different cultures. Michigan, Africa, Brazil, Florida, they are all different. This is why the variable determination department is in existence."

I began to feel that this wasn't going to turn out well at all. Couldn't it be, I suggested, that all people wanted was flypaper that wasn't unattractive, that could be handled easily, and that imprisoned flies once they landed on it? What was so different about that in all those places?

"Oh, the people are no problem," Hal said. "It's the flies. Why should we go to the expense of making 26-pound stickum when 24-pound will do? If we didn't do all that research, we would have to make a consistent type of stickum, a sort of 'world stickum,' and use it everywhere."

I noted that this was what most companies tried to do and that I had never heard of an exterminator trying to please the victim rather than the customer. However, I suggested that since QFC had a lot of research and scientific capacity, it should have those folks take a look at the processes. It just might be possible that doing the process right the first time would eliminate the need for the variation activities.

Hal pointed out to me that the labs were in his department

specifically because some quality control people had tried all this years ago and caused a lot of trouble. The variation department contained about half the people in the company. No one wanted to offend them.

"I'm sorry there is nothing your firm can do for us at this time," he said. "But if you branch out into insect demographics, we would be glad to talk with you or your people."

With that he led me to the door and bowed me out. It was not a new experience for me.

When I was a reliability engineer, I once got all wrapped up in the failure rate of a small relay. It failed about 10 percent of the time, usually after it had been installed on a printed circuit board. Expensive rework was required to remove and replace it. This was a major problem for the assembly area.

I took one of the relays up to the purchasing department to see whether they could help. After a little discussion, I finally found the buyer responsible for the item.

"That relay has an AQL (acceptable quality level) of 2.5 percent," he said. "So it really isn't too far out of line to have a 10 percent rate at final test. However, I will contact the supplier and see whether they can't do better."

"What does an AQL have to do with the rejection rate being all right?" I asked. "The AQL only refers to how much acceptance will be done to it when it comes to us. Why don't we ask these people to send us relays that work?"

"That would cost a fortune," he said. "They would double the price if we asked them to do that."

"Well, it certainly costs us a lot in order to find and fix these things. If the supplier were sincere with us, he wouldn't send us these nonconforming relays."

He looked up at me and spoke the words I have heard many times since then: "Don't you have anything else to do?"

Here is a case where purchasing is using an inspection technique to tell a supplier that it is not necessary to send products that conform to the requirement.

I have fought that for years, as mentioned earlier. Those who expound such methods have usually not worked in business. Reality has a way of making AQLs less practical.

Let me conduct a tour through the development of the concept of Zero Defects. That will provide a much clearer comprehension of what quality is all about.

When I was a line quality manager we had a quality management system that was admired by our customers and our peers. I tell you that not to brag about it, but so you will know that we were not considered to have problems. Everything was running well.

When we began to deliver missiles (and the accompanying paperwork) to Cape Canaveral for test firing, we started to get telexes back.

"The missile had eleven defects," said one. The problems included some serious ones, such as an uncaged gyro, and mostly nonserious ones, such as loose items or scratches.

After the first six missiles had been received, found wanting because of nine to twelve defects, and fixed, they were all successfully test-fired. Everyone was happy with the flights but not so happy with the defects.

"We have to do something about these problems with the missiles going to the Cape," said the GM.

"But there are only a few," I noted. "There must be 50,000 parts in a bird. If each has only six ways of being wrong, that gives us a potential of 300,000 possible errors. We only have ten or so."

I explained the laws of probability. "The normal distribution curve shows that 99.73 percent of related things fall under the curve. There are some people who are 3 feet high and some over 7 feet. The rest of us are inside the limits of the curve."

The GM smiled, told me he was very pleased with my work, and thought that I had been doing the best job of what we already knew how to do that he had ever seen. However, he was serious about the situation.

"Somewhere in the world," he said, "there is a quality manager who can get me products and services with no problems in them. I sure would like it to be you."

I realized suddenly that we were now talking about survival, not about the niceties of quality. (Most companies don't get serious about quality until survival becomes a question.)

As I slogged back to my office I took a detour and walked around the plant site. It was beginning to dawn on me that I was the cause of this problem. With all the complaining I had done about quality levels, I had let defects slip into the system. We actually had a fifteen-defect AQL on birds going to the Cape.

I stopped at the guard gate and called the office to summon all the quality department managers to a meeting. Sixty minutes later all sixty-five of them were gathered in the conference room.

"Ladies and gentlemen," I started, "I have been leading you astray. We are still operating on the idea that errors are inevitable and that our job is to control them as best as possible.

"That is wrong. What we have to do is to concentrate our work on getting things done right the first time. We are going to have a new standard of performance: Zero Defects. Nothing else is acceptable."

There was a moment of silence and a general shrugging of shoulders. The test manager stood up and said, "Phil, this is not a good thing to do. We will look ridiculous to the rest of the company. What do we do when we make an error?"

I smiled at him. "Somewhere in the world," I said, "there is a test manager who can get me hardware with no defects in it."

He sat down and we began.

Two missiles later the people at the Cape called and said that they were unable to find anything wrong with the missile we had just sent over. I asked them to send a telegram, and I still have it.

After that we didn't have much trouble in that area.

However, what we had really shown was that a dedicated group of people, with time and equipment, can find everything that can go wrong with a complicated piece of equipment. That was a good thing to do, but we had to prevent the problems from being there in the first place.

It was not enough to do "super appraisal." That was expensive and impractical. It required a level of examination that was not possible to maintain. We had to learn how to prevent.

So I wrote the concept of Zero Defects. The original concept from 1961 goes like this:

> People are carefully conditioned throughout their private life to accept the fact that people are not perfect and will therefore make mistakes. By the time they seek an industrial life, this belief is firmly rooted. It becomes fashionable to say, "People are humans and humans make mistakes. Nothing can ever be perfect as long as people take part in it," and so it goes.
>
> And people do make mistakes, particularly those who expect to make some each day and do not become upset when they happen. You might say they have accepted a standard that requires a few mistakes in order to be certified as a human.
>
> The question must arise, then, as to whether people have a built-in defect ratio. Do they always make the same percentage of errors in each thing they do? Like cashing their paycheck, for instance.

Can we assume that a person who errs in 5 percent of their industrial activities will be shortchanged on 5 percent of the checks they cash each year? Will they forget to pay their income tax 5 percent of the time? Will they go home to the wrong house several times each month?

If these assumptions are wrong, then errors must be a function of the importance that a person places on specific things. People are more careful about one act than another. They have learned to accept the fact that it is all right to make mistakes at work, but not permissible to defraud the government. In short, a dual attitude has developed. In some things people are willing to accept imperfection; in others the amount of defects must be zero.

Mistakes are caused by two factors: lack of knowledge and lack of attention. Knowledge can be measured and deficiencies corrected through tried-and-true means. Lack of attention must be corrected by the person himself or herself, through an acute reappraisal of his or her moral values. Lack of attention is an attitude problem. The person who commits himself or herself to watch each detail and carefully avoid error takes the giant step toward setting a goal of Zero Defects in all things.

We explained this to the management and then to the employees. Everyone decided to take a whack at it and we began. As we moved along it became necessary to have a strong communication program to let everyone know what was happening. People began to send in ideas, so an error-cause removal system was developed.

The error rates in manufacturing, engineering, purchasing, and other places where measurements were being made dropped 40 percent almost immediately. The quality department began to find less and less in their normal appraisals.

Two or three magazines wrote articles about us. *Time* ran one column on ZD and we were immediately inundated with letters from companies wanting to learn more about this "motivation" program.

I kept telling everyone that ZD was a management performance standard. But my efforts didn't change any minds. Everyone wanted something easy. One person even asked me whether I didn't have anything else to do.

The only people who utilized the concept for what it was were the Japanese Management Association. In cooperation with Nippon Electric Company, they established the concept in Japan. (When I went to Tokyo in 1980 they had a party for me to celebrate sixteen years of ZD in Japan.)

The whole United States, particularly the defense industry, went ZD-happy for two years. Then the "motivation" wore off and they went chasing something else. It has always been a big disappointment to me that the American quality professionals, in particular, never took time to understand ZD. I have found it takes about three minutes to explain Zero Defects to an open mind.

At ITT, beginning in 1965, we installed ZD throughout the world as the performance standard. It worked very well in several languages. People will perform to the standard they are given, provided they understand it. When the standard is wishy-washy like "excellence," "AQL," "pride," or some such, their work varies from day to day. When the standard is specific like Zero Defects, defect-free, or DIRFT (for Do It Right the First Time), people will learn to prevent problems.

## WHY A SPECIFIC PERFORMANCE STANDARD?

All the results in a company are made by people. Each service or product is created by the thousands of tasks that go on in the company and in its dealings with its suppliers. Each of these tasks has to be done properly if the end result is to be what is needed. People have to know that they can depend on each other. A shortstop should be able to know where first base is every day. What one department sends another should be as promised. When this happens, then people can become realistic about the requirements they impose on each other. No one has to ask for something twice as large or as fast as they need just to make certain they get what they really want.

This is the reason for a performance standard that cannot be misunderstood.

The third Absolute of Quality Management is:

THE PERFORMANCE STANDARD MUST BE ZERO DEFECTS, NOT "THAT'S CLOSE ENOUGH."

# 9

# The Fourth Absolute: The Measurement of Quality Is the Price of Nonconformance

The main problem of quality as a management concern is that it is not taught in management schools. It is not considered to be a management function, but rather a technical one. The reason is that quality is never looked at in financial terms the way everything else is. As we noted earlier, it is always thought of in some relative fashion, as in degrees of goodness. However, with the pressure on quality erupting worldwide and the difficulty in getting senior management to do something about it, it becomes apparent that a new measurement is needed for quality. The best measurement for this subject is the same as for any other—money.

The cost of quality has been a subject of discussion for twenty-five years. However, it has only been used as a means of measuring defects on the manufacturing line. It has not been used as a management tool. That's because it hasn't been presented to management in terms it can understand.

Cost of quality is divided into two areas—the price of nonconformance (PONC) and the price of conformance (POC). Prices of nonconformance are all the expenses involved in doing things wrong. This includes the efforts to correct salespersons' orders when they come in, to correct the procedures that are drawn up to implement orders and to correct the product or the service as it goes along, to do work over, and to pay for warranty and other nonconformance claims. When you add all these together it is an enormous amount of money, representing 20 percent or more of sales in manufacturing companies and 35 percent of operating costs in service companies.

Price of conformance is what is necessary to spend to make things come out right. This includes most of the professional quality functions, all prevention efforts, and quality education. It also covers such areas as procedural or product qualification. It usually represents about 3 to 4 percent of sales in a well-run company.

When the comptroller figures out the price of nonconformance—and some help is almost always needed to do this since everybody else is interested in keeping the number very low—he or she can come up with a procedure that can be used forever. Then the price of nonconformance can be used (1) as a whole to track whether the company is improving, and (2) as a basis for finding out where the most lucrative corrective-action opportunities reside.

It soon becomes apparent which products, services, and departments are the main contributors to this number. Most often they do not even know that it is their problem.

The results of most quality improvement systems are measured in terms of indexes or other kind of charts. When shown an index chart on overall quality, executives really don't know what to do about it, because it has no meaning to them. They don't know how to take any action. That is why the quality professionals have never gotten invited to any important meetings over the years.

Collecting the cost of quality is not a difficult task, but it very rarely gets done in a company. The main reason is that those charged with the collecting try to identify each last penny. Then they want to show it in some noncritical way. As a result companies have been working on gathering their cost of quality for a dozen years.

In fact it can be all put together in just a few days. The first analysis may only get 70 or 85 percent of the total, but that will be such an alarming number that you won't really need the rest. Over the years more reveals itself on how to calculate and calibrate the cost. It can be expanded and adjusted as necessary.

I will not go into all the details of putting it together here. That has been documented before, particularly in *Quality Is Free*. However, the rule is: take everything that would not have to be done if everything were done right the first time and count that as the price of nonconformance.

The fourth Absolute of Quality Management is:

THE MEASUREMENT OF QUALITY IS THE PRICE OF NONCONFORMANCE, NOT INDEXES.

# 10
# Education

I've always believed that the standard process of business education does not do a good job of transferring understanding to the student. This is not an original belief; probably few others have ever been satisfied either.

Techniques developed by educators to guarantee that the student will learn a specific bit of information pile up on the rocks of human individuality. There is just no standard way to guarantee comprehension. Too much depends on the student and what he or she, as an individual, wants to learn.

When dealing with a business situation, it is necessary to make certain not only that the material is interesting and is presented in an interesting way, but also that it contains the information actually needed in order to develop things further. It is not enough to make a cookbook for something like hassle elimination, because that will just add another level of hassle to the operation. People will set up procedures to implement the items in the cookbook, and pretty soon compliance with the procedures, or lack of compliance, will become a bigger hassle than it all was in the first place.

Producing a hassle-free company requires the continual transfer of information from person to person. Education in some form has to become routine. Everyone has to have that common language, the skills to do the job, and the understanding of each one's personal role in keeping the wheels of the company moving.

Everyone talks about the need to do things right the first time and no one really wants to do things right the second time. Yet in real life it may be the third time before anything gets done properly. The basic concepts and techniques necessary to eliminate hassle must be given to all the people of the company in a planned fashion.

A solid understanding of a subject means comprehension. I'm always amazed whenever I have the opportunity to chat with one of the touring golf pros to find out how much they know about the game. They understand the way the balls are made, they understand the logic and mathematics of the clubs, they recognize what happens to a stroke under different circumstances. These things are indicative of a really solid, basic comprehension of what goes on inside the game of golf. Quality, like golf, is one of those areas where one can get along for years with just a veneer of information. To make a hassle-free, prevention-oriented company requires that everyone really know what is involved.

The Absolutes of Quality Management must be understood by every single individual. These are the common language of quality.

The fourteen-step process of quality improvement needs to be understood by the management team since they are responsible for making it happen. These steps are all reviewed in Chapter 11; they were covered in great detail in *Quality Is Free.*

The individual's role in causing quality must be understood by each and every person in the company. Those involved in specific functions have to have a special education in order to carry out their roles. Those include supplier quality management, cost of quality determination, and other areas.

The overall educational aspect requires executive education, wherein senior management can learn its role; management education, wherein those who must implement the process learn how to do it; an employee education system, wherein all the employees of the company learn how to comprehend their roles; and workshops, wherein special functions such as purchasing, accounting, quality, marketing, and so forth can learn how to do the individual and special things that are important in their world.

Before we discuss the content of these different courses, it is important to note that one of the best ways of teaching is by means of case histories. However, in any typical class very few people will read the entire case and actually be in a position to discuss and comment on it. So what we have learned to do is dramatize the case by using actors and videos and actually present it as a show. The students have a fact sheet that contains any specific numbers pertinent to the discussion, and then everybody can become involved. (It is also necessary to qualify company managers as instructors for each and every module of the educational process. Instructors who are not qualified and monitored soon go into business for themselves and end up teaching all kinds of strange things that don't apply to what you are after.)

## EXECUTIVE EDUCATION

The purpose of executive education is to help senior people understand their role in causing problems and then causing improvement in the quality process. Executives need to understand what everyone else is going to be taught, they need to understand how they are supposed to react in nonconformance situations, and they need to understand what they can do to encourage the improvement process being implemented. Because they are the overall managers of the company, everything they do is important and watched. Therefore, they need a sound understanding. Executive education requires two and one-half days at a minimum. These must be spent off-site, and the participants must be oriented prior to the session so that they will not be calling back at every break to see whether the company is still there. The course content revolves around the Absolutes of Quality Management, the strategy of quality improvement, the fourteen-step process, the education system, corrective action, and several workshops that ensure understanding of all these subjects and more. Classes contain no more than twenty-two students.

## MANAGEMENT EDUCATION

Management education covers four and one-half days, and each class is limited to no more than twenty-two people. In management education, all the content from executive education is covered with the addition of several items. The fourteen-step process is covered in great detail to ensure that the students have a complete understanding of what goes on in the steps, because behind every action there's an action.

In addition, management education participants need to spend some time on presentation. It is important that this group recognize they have to present the quality improvement case on a continuous basis. So they need to be able to understand some of the requirements of standing on one's feet and making the necessary communications. Because management education ties all the material together with workshops and interplay, it is a much more solid basis for actually doing work.

## EMPLOYEE EDUCATION

The other 95 percent of the people in the company receive their primary quality education from student notebooks. Such a system needs to contain some specific material. We've learned to put together each segment in a standard way. One, material to read prior to coming to class on the subject. Two, a video, usually fifteen minutes, that explains concepts to be discussed during that module. The video is constructed using actors and original scripts. Three, a workshop in which the concept can be applied to something with which the student is familiar. Four, discussion in which the instructor leads the students in determining how the concept applies inside that particular company. Five, a work assignment. The student should take something back to the workplace and apply it to get a firsthand feeling on how the particular concept relates to real life.

These overall segments put consistency into employee education. We wound up with fifteen sessions, each of which covers about two hours and contains all the above. All the sessions are taught by company management, who must be properly trained as instructors. That usually requires two weeks.

## THE SESSIONS

Titles of the sessions and brief descriptions of their content are as follows.

1. *The Need for Quality Improvement.* This session helps people recognize that many customers are disappointed by the lack of conformance in products or services they receive. The company commitment is also shown by a brief videotape of the chief executive.
2. *The Concepts of Quality Improvement.* This is a presentation of the Absolutes of Quality Management in a way that everybody can understand them.
3. *Identification of Requirements.* By means of an input/output analysis of any job, the student is shown that it is possible to identify the requirement components.
4. *Measurement of Conformance.* Taking some of the requirements from Session III, the students learn how to measure how well they are being met. One purpose of this session is to show that measurement is merely a form of communication; it's not something of great concern in a negative manner.
5. *Prevention of Nonconformance.* Having identified a requirement and then measured conformance or lack of it, we discuss the prevention of nonconformance in that area.
6. *The Need for a Performance Standard.* This session emphasizes the necessity of a performance standard that cannot be misunderstood. The workshop in this particular module helps the student learn how to explain Zero Defects to other people.
7. *The Price of Nonconformance.* All the content that goes into nonconformance and conformance is laid out in a manner that the student can use to see what it costs in his or her own area not to conform to the requirements that were identified in Session III.

These first seven sessions are used to provide the basis so that the participant can understand exactly what is involved in causing quality improvement.

The second seven sessions show how all this happens in the company and makes clear where some of the root causes occur. Two fictitious companies are used—the Quality Briefcase Company and the Complete Insurance Company—because many people think that manufacturing and service are different. One of the things that the sessions show is that there is very little difference.

8. *Quality Briefcase Company.* The executives of the company

are interviewed and give their feelings about why problems of quality arise and what needs to be done about them.

9. and 10. *Elimination of Nonconformance (Parts I and II).* These sessions show how the Quality Briefcase Company, having gotten its attitude straight, approaches the problems of nonconformance and their elimination. The coordinator uses the inadequacies of his own golf game in order to demonstrate how problems can be analyzed and identified.

11. *Team Approach to Problem Elimination.* The field service team of the Quality Briefcase Company is used to show that examining the problems together produces more practical solutions.

12. *The Company's Role in Causing Quality Improvement.* The presidents of the two participating companies discuss the differences in their companies and the problems they have in causing quality to happen. This is a very revealing insight for most people.

13. *The Manager's Role in Causing Quality Improvement.* This includes a discussion with an operating manager who is having problems with housekeeping, with getting material into his area, and with getting it out in time. There is a self-test after this, so that all those participating can determine whether they are having a positive influence on their areas.

14. *The Supplier's Role in Causing Quality Improvement.* Here the emphasis is not only on the supplier who brings things in from outside, but also on the internal supplier. Most of us receive material we work on from someone else in the organization. A conversation with some of the participants in the first fourteen sessions follows, and a statement is made by each participant as to what he or she is going to do differently having attended this course.

15. Summary.

The entire education process can be summarized in what I call the "six C's": Comprehension, Commitment, Competence, Communication, Correction, and Continuance.

*Comprehension* is the understanding of what is necessary and the abandonment of the "conventional-wisdom" way of thinking. This is the key to the cultural change required by companies that are determined to improve.

*Commitment* is the expression of dedication on the part of management first and everyone else soon after. It is the deep-seated determination to cause the cultural change. In management's case this is demonstrated by example and by positive thinking.

*Competence* is the implementation of the improvement process in a methodical way. Everything must be dealt with and applied in a way that will cause the cultural alteration to take place. This is no place for manipulation or motivation.

*Communication* is the complete understanding and support of all people in the corporate society including suppliers and customers. It happens only when the company reaches out to them and makes sure they recognize their role in causing quality to happen.

*Correction* is the elimination of the opportunities of error by identifying current problems and tracking them back to their basic cause. It is easy to fix problems, particularly old ones. It requires all of the above C's to eliminate them.

*Continuance* is the unyielding remembrance of how things used to be and how they are going to be. A formal effort is required forever, no matter how well everything is turning out.

## EDUCATION: A COMMENT

While writing this chapter, I could have laid out the entire content of the classes we teach in the Quality College. However, I thought this would imply that we're trying to peddle the college. That would not be true since the college only deals with whole companies, not with individuals.

But the formality of an education structure is important and needs to be discussed—it has a meaning in itself. The content of the courses is important, but it represents only a portion of the educational experience. The classroom environment, the credibility of the instructors, the appearance of the student material—all are part of it. The executive student, in particular, has to be placed in a condition to respond.

All this is true in most management education, but it is particularly true when the subject is quality. The message can be garbled if things don't go just right. Every activity and action on the part of the staff must be Zero Defects in every way. The students become hypersensitive to everything that happens and are very complimentary when right things happen and quite caustic when observing wrong.

Let's list a few of the nonteaching activities that happen in the life of the student and then see what is involved in making these happen.

1. Letter containing information and instructions should be sent to the student to arrive two weeks prior to class.
2. Student should be briefed by company people in order to know what to expect. This provides an opportunity to reaffirm the company commitment.
3. On arrival at the hotel the student should receive a package explaining what will happen over the next few days.
4. The instructor should welcome the class and introduce the staff members they will work with during their stay.
5. Classes should be conducted in one-hour modules. At the end of each hour the class might have a ten-minute period for relaxing and getting to know each other.
6. Case histories should be used in all the classes, as well as in the workshops, in order to provide the maximum learning experience. The case histories can be filmed using actors in lifelike situations. This eliminates the problem of having some people who read case material and some who do not. Everyone can watch the video and relate to it. The cases should be original and not reflect one particular company. This lets specific learning messages get across.
7. Homework should be required each evening in order to help the students keep their minds on the subject. It is important that they become immersed in it. The point is to change a management concept, not teach a bunch of techniques.
8. The final exercise should require the student to respond to a case by supplying a personal solution for part of it. When all the students are from one company the assignment can be given to teams.
9. At graduation students should receive a plaque, a pin, and a few trinkets to help remind them of the time they spent.

The business of making all this happen is a vital part of the education itself. The instructor can use the administration of the class to show that ZD is possible. When something goes wrong students can witness the corrective action.

A few principles are involved in creating the system.

First, it has to be clear that the idea is to not require the student to learn how to obey, for example, by following instructions for checking into a hotel. We do not want to hassle the student.

Second, all the associates involved must recognize how important their personal contribution is to making everything come off in order.

Third, a clear set of requirements, including job descriptions, must be prepared and understood and results measured. There can be no mysterious unknowns.

Fourth, all the outside agencies involved must understand things the way we do. Restaurants, hotels, printers—all have to know their jobs and the way they fit into the entire operation.

Having said all that, let's take a look at some of the special activities that are involved in making it all happen.

1. The letter to the student should be prepared so that it does not give information that is confusing. There can be a different one for each course, perhaps stored in a word processor.
2. We learned early that the students will not know why they are attending the program unless someone tells them. That someone needs to be from their own company. The main message is to reaffirm the management commitment and explain that students are being sent to learn, not to find ways of not doing things.
3. The sessions should be structured to bring the students along a thought process. We have found that one hour is the right amount of time for each part, and some parts may cover several module-length periods. The materials needed for each session should be placed in the room in advance.
4. Classes should start on time. If the instructor begins even if everyone is not there, they get the idea quickly.
5. Case histories provide a basis for discussion, but the students have to know the content. We tried all kinds of ways to get people to read and understand them. We recommend instead the preparation of scripts and films.
6. Homework should be assigned to lead the students into the next day, not to fill them with information at night. We want students to learn so the definitions can be made clear.
7. The preparation "for going home and doing it" consists of hav-

ing the executives stand up and explain "it" to others in the same boat. The presentations should be discussed by the training staff.

There should be a master list for each class that specifies what material is given when, what articles are laid out for presentation, and what instructor speaks for how long and on what. These are the requirements and each job is designed to fulfill them.

The instructors should receive a reminder each day from the senior executive: Be interesting.

# 11
# Implementation

Concepts are essential, and the education to understand them is a must; however, nothing happens unless somebody actively does something. "Doing something" in the case of quality improvement requires that actions be taken to actually change the culture and management style of the company.

This change is *from* being goodness-, appraisal-, quality level–, and index-oriented *to* being conformance-, prevention-, error-free-, and money-oriented.

Every year or so, a new book hits the business community that really turns everyone on. The book describes a system of performance improvement, either personal or corporate, and explains clearly the advantages of managing in that manner. However, even though a lot of executives read it, and even though they sincerely desire to accomplish what is described inside it, very little actually changes. This is not because the concepts or techniques are impractical or illogical. It is not because no one wants the improved results to happen.

The reason very little changes is that implementation is not attacked in a methodical manner as a matter of corporate priority. Somehow management feels that if copies of the book are given to all the members of the management team, and if the subject is discussed, it will just all happen. I hardly ever attend a large function at which an executive doesn't tell me that his or her company bought 4,000 copies of *Quality Is Free* and passed them out—but nothing happened. When a bank decides to expand into branch offices it doesn't just send someone out to buy up a bunch of corner lots. When a retail store chain wants to go into shopping centers it doesn't just pick the one nearest the marketing director's house. When the information system begins to strangle, the company doesn't just call for a bigger computer. For all these actions the same general pattern is followed: an objective is described, a team is put together to design the approach, and the company moves methodically through that process. Measurements are taken all along the way and course corrections are made as required. All the employees who need to know are taught until all of them understand. Then the ones who didn't need to know then, but do need to know now, are taught. A routine instruction operation is set up to make certain that all new employees, regardless of level, achieve the same comprehension as everyone else. That way the subject can be discussed.

The process of installing quality improvement is a journey that never ends. Changing a culture so that it never slips back is not something that is accomplished quickly. Nothing happens just because it is the best thing to do, or just because it is worthwhile.

Not everyone is in love with the new communications systems. Not every manager wants a computer terminal on his or her desk. Some people still like to lick a stamp and place it on a letter or to bring the old-fashioned lunch pail to the office instead of a brown bag in a briefcase.

Changing a culture is not a matter of teaching people a bunch of new techniques, or replacing their behavior patterns with new ones. It is a matter of exchanging values and providing role models. This is done by changing attitudes.

Television commercials showing people of different races having easy relationships have done more to eliminate bigotry than all the laws and sermons ever produced. To see people of different races treat each other like equals who have similar problems and feelings is very effective. Films and TV shows with women performing executive tasks and being competent in the professions have given more credibility to woman's liberation than all the speeches and articles put together. The reality of these situations never occurs to some people unless they witness them and see their acceptance by others.

Causing quality improvement to be part of the corporate cultural woodwork entails much the same process. All those terrible things that people are doing, the hassles that cause defects, are done with the best of intentions. All of us are doing what we think the company wants. We are doing what it has told us to do by the way it acts and by the scenes it shows us.

The culture we have now was caused. We don't need to know how it all happened and we don't need to condemn it. We don't need to define the differences between what now exists and what we would like to have. All we have to do is describe the future as we would like it to be and then march on down the yellow brick road.

That is the road that we have to lay out—the path of our journey. It is made up of fourteen things that we have to do something about in a certain order. Then we lay out the same fourteen blocks again and keep building. The blocks get more familiar and comfortable as we repeat, but they provide the foundation forever.

Although I have been working with these blocks for twenty years I still learn about them every day. They sound like they are not complex, but they are. Some look like they can be ignored, but they cannot. Let's take a look inside them and see what goes on. (The content of the steps was described in *Quality Is Free*. Here we can discuss the experience we have gained by helping dozens of companies implement them.)

The fourteen steps of quality improvement are:

1. Management commitment
2. Quality improvement team
3. Measurement
4. Cost of quality
5. Quality awareness
6. Corrective action
7. ZD planning
8. Employee education
9. ZD Day
10. Goal setting
11. Error-cause removal
12. Recognition
13. Quality councils
14. Do it over again

How do companies get involved in quality improvement?

The pattern we see happening at this writing is that top management asks a senior executive to take a look at "this business of quality." The assignment includes finding out what is happening in other companies, figuring out what should be done about it, and then getting everything moving.

It is always a very positive step with remarkably little preconception imbedded in it. This is a sensitivity that is new since 1980. Prior to that time most top managers I met had their own firm ideas on what made quality happen and not happen. Even today some are out making speeches about technique-oriented solutions without recognizing that very little has changed in their companies as a result of those techniques.

When the investigation of the "quality business" is complete, the executive is usually faced with a dilemma: how to bell the cat? The very ones who established the mission are those who are going to have to be corrected first. They are going to have to be educated and then participate in telling everyone that they have changed their ways.

In talking with companies who are experiencing improvement the executive has realized that senior management is the key to the solution, as well as the cause of the problem. It takes a little longer to explain that than one sentence, but it does get explained. Only when the management team becomes educated and sets out on its mission of changing the culture of the company can it hope to reap the rewards such a change produces.

However, it will soon become apparent that not everyone believes in top management's new policy. Its credibility is low. It is well known to get all fired up about some new program and then switch it off after everyone is working away on the subject. The problem used to be: "How do we get management interested enough in quality to do something about it?" Now the problem is: "How do we get the people to believe that we are really going to do something and stick with it?"

The culture of the company is going to change only when all employees absorb the common language of quality and begin to understand their individual roles in making quality improvement happen. As their education takes effect, the employees will expect management to put quality first among equals. From that point on they expect to see management insisting on conformance to requirements and they expect to get used to it.

This is where the yellow brick road begins. Let's take the fourteen items one at a time. I will cover each one in some detail, because we know a great deal more about them now after several years of experience.

# STEP 1: MANAGEMENT COMMITMENT

When executives of a company first come to the Quality College they always ask the same question: "Is our senior management really committed to this process?"

We remind them that those individuals have already come to the school and show them the photo of their class. This doesn't satisfy their question. Then it is proper to ask exactly what senior management would have to do in order to prove that it was serious. This places the commitment in perspective. Obviously some faith in management is required, as well as some actions that will support that faith. The way senior management acts toward accepting anything less than the requirements lays the foundation of this faith.

A few actions are necessary. First, a corporate policy on quality needs to be issued. This policy should make it clear that the commitment is real and understandable. It must have no weasel words in it.

Second, quality should be made the first item on the agenda of the regular management status meetings. It should come before finance and be discussed in specific terms.

Third, the CEO and the COO need to compose clear quality speeches in their minds and, as they go around the company, deliver them to everyone whose path they cross.

Let's take these items in turn.

## The Company Quality Policy

Writing the policy is the first act; getting everyone to understand that it exists and is a serious message is the second act. We recommend that the policy say this: "We will deliver defect-free products and services to our clients, on time."

However, each company has its own way of communicating such information. To demonstrate this we have framed the policies of many companies and hung them around the Quality College. I have listed some of them below to show how this basic thought can be said in different ways. The only rule is that the policy should be so clear that it cannot be misunderstood.

### 3M COMMERCIAL TAPE DIVISION

3M Commercial Tape Division is committed to a policy of conformance to requirements for each function of the organization and for customer satisfaction, or we will change the requirements to what we and our customers really need.

Implementation of this policy makes it essential that each person be committed to performance exactly as required.

It is our basic operating philosophy to concentrate on prevention methods to make quality a way of life and perpetuate an attitude of "Do It Right the First Time."

### IBM
### RESEARCH TRIANGLE PARK, RALEIGH

We will deliver defect-free competitive products and services on time to our customers.

### ARMCO, INC.
### MIDWESTERN STEEL DIVISION

The policy of the Midwestern Steel Division of Armco is to provide products that conform to our customers' requirements and deliver them on time and at a competitive price. Our name must represent quality to our vendors, ourselves, and our customers.

### BURROUGHS

We shall strive for excellence in all endeavors.

We shall set our goals to achieve total customer satisfaction and to deliver error-free competitive products on time, with service second to none.

### BECHTEL
### ANN ARBOR POWER DIVISION

In order to improve quality we shall provide clearly stated requirements, expecting each person to do the job right the first time, in accordance with those requirements or cause the requirements to be officially changed.

### MILLIKEN

Milliken and Company is dedicated to providing products and services designed to be at that level of quality which will best help its customers to grow and prosper. Its operational area (Research and

Development, Marketing, Manufacturing, Administrative, Service) will be expected to perform its functions exactly as written in carefully prepared specifications.

## PHILIP CROSBY ASSOCIATES

We will perform defect-free work for our clients and our associates. We will fully understand the requirements for our jobs and the systems that support us. We will conform to *those requirements at all times*.

After the policy has been made clear, a usable quality status report must be brought to the important meetings. Quality should be listed as a CEO agenda item in order to reaffirm its seriousness. Reports will' be made on the status of:

- The quality improvement process
  - How many employees have been educated?
  - Are the teams functioning properly?
  - What success stories do we have to share?
  - What problems need action?
- The cost of quality
  - Do we have the format in all operations?
  - What are the trends?
  - Where do the biggest improvement opportunities lie?
  - What problems need action?
- Conformance
  - Are we meeting our requirements?
  - What actions do we have to take to emphasize the need to meet them?

In order to continually reaffirm dedication to the process, the CEO will have to get used to making a short speech regularly. Here is a useful version:

## CEO SPEECH

I recognize that quality is a very popular subject today. Our customers are concerned about it, the nation is concerned, we as a company and as individuals are concerned.

We have done a great deal of investigation on the subject and have learned quite a bit. One thing that has come through very clearly is that there are no simple, painless cures for solving the problem of quality. It takes determination, education, and then a clear process of implementation.

Our studies show that a company gains improvement quickly when it takes quality seriously. They also show that it takes several years before it all becomes part of the routine, if it ever does.

We have also learned that quality begins with senior management. So every member of management is attending special classes to help us develop a common language of quality and recognize what our individual role is. Everyone in the company will attend classes on this subject.

We are going to deliver defect-free products and services to our customers. After all, this is what we have told them we will do in our advertisements and discussions. There is no more important task for our team to accomplish.

The company that delivers what it says it will deliver is the company that will lead its industry.

We have the people, the equipment, and the knowledge to make it all happen. We are committed to quality forever.

The senior executives will have to deliver this type of meditation for the rest of their days. They will also have to face the problem of, "Is this good enough?" Everyone wants to make absolutely certain that minds have not been changed. And reassurance is demanded on a daily basis until confidence grows.

About six months after we started Zero Defects in 1962, a top manager came to me with a nonconforming assembly. It was in good shape electronically, but the case had a gouge in it.

"I have seen engineering and they say that this mark doesn't affect form, fit, or function. Marketing says the customer doesn't mind because they just bury the unit anyway. We can't rework it and it would cost $35,000 to make new cases. We will only do twenty-three of these units and they are already all made."

"So what do you want from me?" I asked.

"Sign it off so we can use it. Your people won't touch it, what with all the ZD emphasis and all."

"You have no problem with me," I replied. "Just get a drawing change notice, tell us where you want the gouge, and we will make certain that all of them come out that way."

The message has to be clear from top to bottom. "We will take the requirements very seriously. If we don't need something, then let's officially change the requirement. But please don't ask me to agree to deviations. We need to spend our time learning how to make things right."

The first time the boss agrees to make an exception, everyone in the company will know about it. However, it works the other way also: they will learn immediately that current urgency didn't override the decision to do what we know we should do.

Management commitment is tested and tested until it can be assumed.

# 12
# Team
# Actions

## STEP 2: THE QUALITY IMPROVEMENT TEAM

The quality improvement team requires a clear direction and leadership. Otherwise people can get so involved with strategy, and the selection of the team, that they forget what the team is for. We need a few rules about what the team does and how its members should be selected. The team is one of the key parts of the process.

The purpose of a team is to guide the process and help it along. It is not to clear each action beforehand, to be the all-wise oracle, or to hold things back. It is to coordinate and support.

The quality improvement team should be made up of individuals who can clear roadblocks for those who want to improve. They should represent the company with the outside world, schedule the education program, and create the companywide events. The team should also represent all functions of the operation. Members should be able to commit those functions without running back to check each time.

The chairperson of the team should be someone with an easy conversational access to the very top management. The chairperson has to understand the overall strategy and have a strong hand in order to alter it if necessary.

The overall process needs a full-time coordinator. The coordinator leads the effort by making things happen and by being certain that the chairperson and the team are all together.

Top management, the coordinator, and the team chairperson lay out the overall strategy, usually with the help of a consultant. This strategy is agreed upon with the team and varied as necessary to meet the very practical needs of the team members.

The team members all need to have the same educational base concerning the quality improvement process or it will never get out of the chute. Those who don't understand the concepts will lead the entire effort into a low-level motivation program. That is where the interest in programs such as quality circles arises.

The team has to understand that we are after change in the attitudes and practices of the supervision of the company, not of the troops. Their turn will come.

## The Purpose of Teams

Quality improvement teams are usually set up with the thought that they will take on the task of "improving things around here." That is not an unrealistic thought.

However, the way in which teams cause the improvement may not be obvious. It is not just through the methodical creation of procedures and actions. Some of this will happen and needs to. The team will also see to the setting up of education activities for all hands. Nothing is more important than that.

But the real learning comes from the experiences that the team members themselves have. After being in hand-to-hand combat with quality improvement for a while, they begin to understand what it is all about. They begin to recognize that it is a very deep and subtle subject. They also begin to enjoy the struggle—and the victories.

The more (nonconflicting) teams that are established at managerial levels, the more people will finally get the comprehension necessary to dehassle the company. The best training is playing.

Consider Little League baseball. Would the purpose of that effort be just to score runs? Is the measurement of success the amount of plate crossings or error-free fieldings that occur? The purpose of childhood athletics is to help the individuals learn more about getting along with others, while understanding themselves better. Comprehension of life is the purpose. Those who expect quality improvement teams to become corrective-action functions have the same limited approach to management as those who count Little League only by the pennants that are won.

Every person who spends time on a quality improvement management team will grow in his or her value to the company—and to himself or herself.

## STEP 3: MEASUREMENT

Many quality improvement teams and, in fact, many companies are very tentative about measurement. They look on it as the ultimate hassle. However, the hassle comes from not having clear measurements. It's when no one can tell how well you're doing that you get frustrated.

Measurement is a normal thing; we are surrounded at all times by measurements. We have clocks, calendars, speedometers, ages, heights, weights, blood pressures, all these things. It is very difficult to have any kind of conversation about anything that doesn't include several measurements. If this were not so, we would be unable to communicate in definite terms.

Measurement is just the habit of seeing how we're going along. One company ran into problems with its management people, who said that there was no way that their jobs could be measured. So the quality improvement team issued each manager a measurement chart of a standard size and color, and a crayon for marking, and asked each manager to think of one thing that could be measured and hang it on the office door. There was a long period of silence for several hours, and then one manager put, "Getting to meetings on time" out there. Soon another one put, "Articles of mail left over at end of day." Another put, "Times secretary needs to find me and cannot find me." Soon everybody had a measurement.

Quality improvement teams struggle around this subject quite a bit until it finally dawns on them that it is not up to them to determine these measurements. All work is a process; you can identify the inputs to work whether you are a bank teller, a cement pourer, or a computer programmer. You receive inputs to your work from other people, other functions, other suppliers. Then you apply your process to it. Your job changes that input in some way, and that results in the output. So you've got input, process, and output.

Each of these lends itself to measurement, and any job can be measured by using that simple pattern. We find that once working people at any level understand this, they can easily create measurements for themselves and help others create measurements. If all else fails, go to the people who receive your output and ask them to tell you how you are doing. The answer is a measurement.

Many managers are concerned about the application of the quality improvement process to the developmental areas. They are really talking about what they see as possible interference with "the creative process." This is a legitimate fear, since many people view creativity as an unmeasurable thing. They're also concerned that many times requirements are ill-defined and probably not finished. This conflicts with the first Absolute of Quality Management. The comprehension problem here is that R&D is really not that much different from any other function. Creativity exists everywhere; it just happens to be formalized there. Requirements that are not complete or final enough also exist everywhere.

Everyone thinks the other people have clear requirements and they alone are left in the dark. As you know, this is not true. The creative person needs to communicate the creation to other people. This can only be done through processes, procedures, and measurements. For instance, take the matter of publishing a book. If authors want people to be able to read their ideas, comprehend them, and do something about them, they are forced to follow certain rules and regulations: they must write in a language the reader can understand, they must use correct sentence structure, they must write in words that are within the reader's comprehension level, and they must organize it all in a manner that follows a logical process. The book itself must be published in a format that can be handled and sold. An index is required and must appear in the back of the book. A table of contents must be in the front of the book. All these are requirements to be met if the idea is to have an opportunity to be heard.

The process has nothing to do with the goodness of the idea. It takes just as much trouble to explain a bad idea as a good one.

## STEP 4: THE COST OF QUALITY

The concept of the cost of quality was discussed in detail under the fourth Absolute of Quality Management. But what we learned about in working with quality improvement teams was that not everybody is in favor of calculating this expense. The quality function inevitably wanted to keep it small, and the nonmanufacturing functions tended to want it limited to the manufacturing costs.

As a result we learned to set up a special workshop to bring together the comptroller and several other interested parties and help them work out a procedure that fit their accounting system. This lets the number be calculated objectively. By supplying this outline to other operating functions, we began to see a pattern whereby everybody could calculate a cost of quality in the standard format and the comptroller could bring them all together for the company.

Many chief executives immediately pounce on the cost of quality as yet another way of measuring the performance of their executives. And once in a while somebody would say, "I want the cost of quality cut 10 percent in this operation within the next two months." Then the CEO would find out that there was nothing difficult about that—all it took was eliminating all the quality department functions. Of course, that is counterproductive and causes a lot more trouble than it saves money. So we learned how to explain to the chief executive and the other senior executives that the cost of quality is a flow. One day they can indeed get around to using it as an executive performance measurement, but in the first segment it has to be treated as a positive rather than a threatening item.

When a company cost of quality has been identified and fed into the regular management process, it serves as a very good and positive stimulus for the quality improvement process itself. There's nothing like money to get management's attention.

The cost of quality has to be pulled together formally and objectively. In this way it can describe trends, because it measures the same things all the time.

## STEP 5: QUALITY AWARENESS

Communications inside corporations and organizations are still a difficult matter. There are so many things to communicate and so many things that people want to know. It is always difficult to make certain that the things it is necessary to communicate are being understood. Many companies try to explain quality and to make people aware of it by setting up supernumerary publications and information systems. This is useful. However, the most effective quality awareness systems seem to be those that use existing systems inside a company. Instead of being in a separate quality newsletter, for instance, quality awareness becomes part of the regular company newsletter. If the company doesn't have an overall communications system, it should start one; every company needs to communicate.

The word "quality" needs to be spread around. People need to be reminded of it. Posters are not naive or showy or unprofessional. One of the most effective means of reminding people turns out to be floor mats. Milliken Company designs and produces floor mats to say things like, "Do It Right the First Time," and all sorts of other quality slogans. Lay these anywhere around the company and people will be reminded. Studies show that for some reason the floor mat causes people to remember more than the posters.

Awareness must be adapted to the culture of the company. It has to fit in. But by the same token it should have a sense of urgency and an individuality about it. People need to know about the management commitment. They need to know about the policy. They need to know about the costs of doing things wrong.

Awareness is not just making publications and promotions and so forth; it is spreading information. One management team I saw was concerned about the errors in computer programming. The team members felt that these put an unusually heavy burden on the mainframe computer, since it was being used primarily for troubleshooting, and they estimated a cost of $250,000 due to the problem of the software's not being close enough to specification before troubleshooting began. Rather than just telling people about this, they borrowed ten brand-new Cadillacs and lined them all up in the front yard. Then they invited everybody out to "see what troubleshooting costs us." That made a big impression.

Awareness of quality extends all the way through management's actions. The way management people talk about quality is important. When quality as conformance to requirements becomes part of the lexicon of the company, then it begins to take effect.

## STEP 6: CORRECTIVE ACTION

Most companies feel they have a corrective action system, yet they still have a lot of problems that don't seem to get solved in any reasonable length of time. I always like to ask audiences, "How many of you have a problem you have been trying to get fixed for two years." There are always many hands raised. This means that some problem has not been identified to the proper people for elimination.

Sometimes people work on the wrong things. Football teams provide an endless source of analogies with which to make a point. Let's assume a hapless mythical team that used a misguided failure-cause analysis as a method of starting corrective action.

Coach Smedley conducted his evaluation after losing the first three games by a score of 14 to 13. He reasoned that the problem was that the opposition had blocked one extra-point try while his team had blocked none. Had his team blocked two, the game would have been won 13 to 12.

The team set out on an intensive practice schedule devoted exclusively to the art of blocking extra points. They practiced all week, set goals for themselves, conducted motivation meetings to keep themselves pumped up. Two games later they blocked 13 extra points. They celebrated the achievement of their target and then went on to deny their opponents dozens of extra points during the remainder of the season. They had quite forgotten that there is more to the game.

Strangely enough, the biggest problem with corrective action is a misunderstanding about what the phrase means. We see corrective action systems established whose sole purpose is to return nonconforming items to a proper conforming status—erroneous computer runs are reprogrammed and replaced, credit card malfunctions are corrected, undersized holes are reamed out to become the proper size. All these are done with a great deal of intensity. We also see that the after-delivery operations are treated merely as maintenance, as fixing, and there's no intensity about getting feedback.

The real purpose of corrective action is to identify and eliminate problems forever. There are several phases. For instance, if you suddenly come face-to-face with a grizzly bear in your backyard, the immediate corrective action is to separate yourself from the bear as quickly as possible and doing nothing to irritate the creature in the meantime. This is a case of replacing nonconformance with conformance and is what most people think corrective action is all about. The bear leaving the woods for your yard is an act of nonconformance. But a discussion at this time is probably not an effective idea.

However, the real problem is, what is the grizzly bear doing in your backyard? In order to answer that, we must determine the cause of the bear's actions and identify what would be necessary in order to keep them from happening again. Perhaps the bear's other sources of food are being jeopardized by something. Perhaps the bear was driven there by some other force. An analysis has to show what actions are necessary in order to keep the bear back in its own part of the world.

The answer would not be to set up an armed camp to protect yourself from the bear. This is the sort of action that takes place when parts of an organization are given the shoot-to-kill license. All that results in is a lot of yard that can't be used and several dead bears.

Corrective action systems have to be based on data that show what the problems are and analyses that show the causes of the problems. Once the root cause has been determined, it can be eliminated. That is what all the corrective action is all about.

One major corrective action step that has become very popular and very useful involves supplier quality management. By identifying the suppliers who are the primary sources of difficulties and conducting meetings with them in a planned fashion, companies can reduce and eliminate the nonconforming services and products that they supply. Half this elimination occurs through nothing more than talking and coming to an understanding about what the product is supposed to be. Thus taking corrective action with suppliers becomes a way of helping identify the requirements, getting them clear, and resolving how they are going to be met and measured. There are very few new causes for error in the supplier-purchaser relationship.

# 13
# Team Executions

## STEP 7: ZERO DEFECTS PLANNING

In many companies, when Zero Defects planning is undertaken the team gets very nervous because they start thinking about having a Zero Defects Day. They think this is going to entail a lot of embarrassment because they are going to have to have a band, straw hats, and balloons and do all kinds of funny things. That is not necessarily the case. ZD is a celebration and will take care of itself. What is important is to know when to do it. We see many companies who want to rush into ZD Day as quickly as possible because they figure that this is going to get them the most rewards. However, there is no need to hurry or to plan for a Zero Defects Day sooner than a year and a half after the beginning of the process.

The Zero Defects commitment represents a major step forward in the thrust and longevity of the quality management process. It should be taken very seriously and planned in a very dignified way, with no hype about it. It is a time for communication, and it needs to be different from all other communications so that it can be recognized. We have participated in dozens of ZD Days as guests and always found them to be delightful occasions.

In planning, the team should take into consideration the speakers involved. Some should represent the customer if there is a predominate customer, some should represent the union if a union is involved, some should represent a city or county faction to show that people outside the company are interested in the activity and the result of it.

## STEP 8: EMPLOYEE EDUCATION

When management finally understands the four Absolutes of Quality Management and starts following the yellow brick road, it comes across the need to educate all the employees in the company. Traditionally this is done by having the training department put together some information, work with a consultant, and come up with a program. However, we have learned in the last few years, among other things, that when this happens it doesn't take very long before what is being taught becomes different from what everyone thinks is being taught. This occurs because of the normal changes in personnel and the normal desire of instructors to make the classes more interesting. Very soon conventional wisdom rears its head, and you have what I call the "lost civilization" experience.

A woman was teaching her son to cook a canned ham. She opened the can, took out the ham, and sliced 3 inches off the end of it. The son asked why she did that and the mother said, "That's the way we've always done it. Ask your grandmother." So the son asked the grandmother, and grandmother replied, "Well, I always cut 3 inches off the end because the roasting pan was very small and it wouldn't fit."

This is the "lost civilization" type of teaching. Things are passed on without anyone's really understanding them.

As a result of having this experience in several areas, we began to realize that the quality improvement process needed more help. So we developed a complete quality education system that would provide a standard message and could be taught by anyone who was trained to use it. The entire system requires thirty hours of class time, plus assignments back in the workplace. Each session consists of a video that explains the concept of that session, a workshop to bring the subject home to the student, and then a discussion to personalize the subject to that company. In addition, there are reading assignments before and after. Then an assignment is taken back to the workplace and reported on at the beginning of the next class.

This required an investment of time and money by the organizations involved in quality improvement, but it resulted in quantum leaps.in improvement. Suddenly everybody understood what quality was about, and they understood that they had to do something about it. It was not something to be done just by the company. People became less tolerant of hassle at all levels of the company as well, and it began to disappear.

## STEP 9: ZERO DEFECTS DAY

There are still those in the quality profession and other isolated areas who think the purpose of Zero Defects Day is to get all the employees together so they can sign a commitment to improve. That does happen, but it's not what the day is all about. Zero Defects Day is a day to get management to stand up and make its commitment in front of everybody in a way it must abide by. It is a time to show all the people, face to face, that management is serious. Many companies are now happily celebrating their third, fourth, and fifth anniversaries of ZD Day. They find it to be a day of rededication that happens every year, a day in which nothing negative happens in the whole company.

Those of us who travel a lot and spend time with people in many fields forget that ours is not a normal pattern of life. A great many people very rarely have exciting days at work. To the overwhelming majority, a well-planned, dignified Zero Defects Day on which management understands what it's talking about is a delight that will be remembered forever.

## STEP 10: GOAL SETTING

Goal setting is something that happens automatically right after measurement. Some quality improvement teams feel that they have to do steps 1 through 14 consecutively, but actually most all of them run in parallel. You never get done with quality education, for instance. The first six steps are all done by management and need to be done first. But when you start measurement, people immediately begin thinking about goals. The ultimate goal, of course, is

Zero Defects, and that is what everybody is after. In the interim, however, intermediate goals move you in that direction.

Goals should be chosen by the group as much as possible and should be put up on a chart for everyone to see. Minor goals should not be accepted.

## STEP 11: ERROR-CAUSE REMOVAL

Error-cause removal is asking people to state the problems they have so that something can be done about them. It is not a suggestion system in which people have to come up with the answer. However, most problem statements contain suggestions that in turn help resolve the problems.

Teams are usually overwhelmed by the number of error causes they receive. The team must ask itself, What are we going to do when we receive an error cause? How are we going to tell the person we received it? How are we going to get it analyzed and acted upon? How are we going to tell the person we did something about it?

This is a more valuable part of the quality improvement process than many people realize, because it is a communication that is peculiar to this system. Employees really love it, and they will respond to it.

## STEP 12: RECOGNITION

Employee recognition is something that is specific to each company, and in *Quality Is Free* we talked about it in great detail. In our company, Philip Crosby Associates (PCA), we set up a Beacon of Quality award. A beacon is a reference point, an unmistakable guideline to keep something heading in the proper direction. This is well understood in navigation of airplanes, ships, and the like. But how about people? Where do we get our beacons when it comes to people?

Each of us sights on some other person, consciously or unconsciously, as a reference point. We have personal mentors if we are lucky, and they serve as examples and reference bases. Most of our understanding about how things should be approached comes from watching individuals we consider "good examples": people who do the right things without making a big deal out of it. In PCA we ask our employees to identify the individuals they feel serve as beacons when it comes to quality. "What person do you see as the standard for quality performance?" The awards each year are planned to coincide with the company's annual black-tie "picnic" held at the Citrus Club in Orlando. Forms are sent to each employee with the request that they nominate the people they feel are qualified. (The only person not eligible is the chairman.) Completed entries are sent to our accountants, Ernst and Whinney for tabulation. Just like on television, on the appointed night the Ernst and Whinney account executive steps forward to hand the chairman the envelope. There are three winners, who are presented with large brass candelabras properly engraved with their names and the title "PCA Beacon of Quality."

Those who received awards were deeply moved by being so chosen by their peers. They would not have been so moved, nor would the rest of the people, had management made the choices. What effect does this award have? First, it recognizes some hard-working, valuable people. Second, it provides a very clear description of what quality performance is. Third, it provides the organization with living, talking, visible beacons of quality that we can all try to emulate on a daily basis.

Very few companies recognize their good performers. Many managers feel, somewhat cynically, that people are being paid to do their jobs and that's that. This attitude reflects an insensitivity to people that is a trademark of many hockey-style managers. It is immature. The creation and development of a recognition program for executives and employees alike is a very important part of quality improvement. It is as important as supplier quality management or supervisor training or cost of quality discovery or the chief executive's comprehension of what quality really means and how to get it.

People don't work for companies; they work for people. Those who don't work well or efficiently need a beacon to line themselves up with. They need to know what is a proper job performance. Our biggest lesson of the past few years has been that many companies rush too quickly into recognition. There is a gratification in giving out trinkets and certificates that makes everybody want to do it. However, the recognition process needs to be thought out and it needs to be done on several levels. I have become even more convinced that money is a very bad form of recognition. It is just not personal enough.

## STEP 13: QUALITY COUNCILS

The idea of quality councils is to bring the quality professionals together and let them learn from each other. They can also support the quality improvement process. There seems to be a very clear demarcation in companies when it comes to the way the quality professionals treat all this new attention to quality. The quality professionals are either actively involved in helping the company eliminate hassle and believe that Zero Defects is indeed achievable, or they are trying to encourage more and more worker motivation and communication programs and to convince top management that Zero Defects is not really obtainable.

## STEP 14: DO IT ALL OVER AGAIN

Time and again we have seen the quality improvement team, after two years of activity or less, turn all its responsibilities over to a brand-new team with perhaps one continuing member. Inevitably the retiring team feels that there is very little left to do. However, its members are always surprised and pleased to see that the new team immediately takes off on many new tacks, develops many new ways of doing things, and causes even more improvement than happened the first time. This is all a result of learning, and of watching and participating. As quality improvement becomes more and more an enduring way of life, as it becomes the culture of the company, the process gains speed and permanence.

This means that the management of a hotel-type company has to do a very intensive job of telling the employees what is expected in terms of performance. Education has to be a vital part of the relationship. Any employee may be called upon to represent the hotel at any moment.

The foundry people need education also, but a smaller portion require the knowledge of how to deal with the customer. It is this difference the people refer to when they break down the world of business into "service" and "manufacturing." When executives come to the Quality College they instinctively say, "Okay, you people are so smart, let's see how you run your company."

They talk to all the associates they can find, poke around in the stockrooms, and, in general, check us out. This is a big help to us because it forces us to take the education of all associates (every employee is an associate) seriously.

In order to have meaningful education, it is necessary to have something to educate people about. That something has to be specific. It has to be based on the job that the associate is doing so that it will help ensure proper performance. However, the education also has to be general so the associate knows what is going on elsewhere and has some information to grow on.

In our service company, the people at the bottom of the organization meet the College participants in the performance of their duties, setting up classrooms, arranging lunches, picking up classes at the motel, preparing coffee for breaks, going to lunch with the class now and then. The instructors come from all the professional levels.

In our manufacturing company, the participants rarely see those who are assembling notebooks, cleaning classrooms, delivering mail, sending out invoices, and so forth.

Everyone has to be prepared for whatever happens. Requirements are clearly defined, job descriptions are prepared by those doing the job, and everyone has to attend an Executive College. Everyone also went to the Dale Carnegie class. It was a big help to each attendee. It was a revelation to all that they could actually stand up and say something. Many have continued on to Toastmaster and Toastmistress.

The difference between manufacturing and service is whether the product finally does something for the customer or an individual does.

As people become involved in the quality improvement process they wonder about the difference between service operations and manufacturing operations. The following story may help in understanding.

# SERVICE VS. MANUFACTURING

"All this stuff is okay for manufacturing, but how about service quality? That is an entirely different thing altogether."

"It is?"

"Certainly it is. When you start talking about service you are in a whole different field. Banks, insurance companies, credit cards, hotels—they are all service operations.

"What I want to know is, do all the things we have been reading to this point apply to service as well as to manufacturing? I suspect they don't. Do we have to have new concepts and techniques?"

"I'm not sure. Perhaps we could do a little analysis together. Give me the name of a service company."

"Okay. How about the American National Bank? It has 35,000 employees and branches all over the world, and it is in the service business."

"Good. Now give me the name of a manufacturing company."

"Entire Motors is a good one. It makes automobiles, trucks, and other vehicles in a lot of places. Both of these are good examples."

"I agree. Now let's take the bank first. What do its people do?"

"They service the accounts of the customers. When people come into the bank to make a deposit or a withdrawal or to borrow some money, the bank's people perform the service for them."

"What happens after the customer leaves?"

"What do you mean, what happens after the customer leaves? Another customer comes along."

"I mean after the deposit, withdrawal, or loan is made—what happens as a result of the transaction? There has to be some paper or computer entry or something involved."

"Oh, yes. Well, the record of the transaction is sent to the recording operation, where the customer's account is adjusted and the necessary papers are prepared in the case of a loan.

"Then the transactions have to be noted in the proper places so the bank's books will balance."

"And all that is done by people? I realize that they use computers, but the people run things?"

"Certainly, and there are a lot of them doing this sort of thing."

"What percentage of the bank's employees are involved in these procedural actions? Would you call these procedural actions?"

"Yes, I would think so. The employees have to follow a procedure to make certain that the actions all get credited to the right place. There is an incredible amount of paperwork involved in banking.

"From what my banking friends tell me, about two-thirds of the people in a bank are involved in these 'back room' operations."

"So they don't serve the customers directly?"

"No, unless they have a problem, in which case they call or write to the customer. But for the most part the customer sees the tellers, some officers, and up-front service people."

"So two-thirds of the bank's employees never have face-to-face contact with the customers, although some of them have contact by phone or letter. And those who work in the bank do their work to agreed-upon procedures and follow prescribed rules of application for their work. Is that correct?"

"Correct. That is the way service companies work."

"Good. Now let's talk about the automobile company. Why do you consider it a manufacturing company?"

"Well, it's obvious. Entire makes cars for people. Every day it buys parts, brings them in, and assembles them into automobiles that customers purchase and use. That is manufacturing."

"Do you have a car?"

"We have two of them."

"Have you ever met an active automobile assembly worker?"

"You mean actually dealt with someone who was putting a car together? No, but I have bought several cars."

"How did you go about it?"

"I went to the dealer's showroom and told the dealer what I wanted. We agreed on a price, and the dealer gave me the car."

"What happened to the paperwork involved?"

"It went through the sales office and I guess to the auto company's office. They produce several million cars a year, so they must have a lot of paper."

"They do a lot of service, too. I did some checking on that company. One-third of their employees are involved in the manufacturing operation. The other two-thirds are selling, servicing, handling paperwork, administering, managing, or doing something that doesn't involve assembling cars."

"What's your message?"

"Well, I am having a hard time determining the difference between these two companies. It looks to me like they both have factories where the product is produced (the 'back room' for one and the assembly line for the other). They both have people who deal with the customers, who sell, who do administration, who handle complaints, who manage, and, in general, who do much the same thing.

"To me there is little difference between someone who inserts six spark plugs in an engine, and someone who separates loan applications by putting them in different-colored folders. They are both performing a service."

"But manufacturing is a technical operation that revolves around machines and systems, while service is a human thing that involves a lot of variables. They are completely different, completely."

"The automobile company has 345,000 employees. Aren't they human? Very few of them actually touch the manufacturing machines and those who do are performing a service. They are not a natural resource; they are pulling levers and pushing buttons and taking measurements and things like that."

"In the service companies they perform services; they don't make anything."

"How many meals does a big hotel serve every day? Where do those meals come from? Is a kitchen a factory?

"When you receive your report from the bank each month, it comes with a printed statement and a pile of checks all arranged in an envelope. How do they get together?"

"They are put together by a bank employee performing a service for me."

"And how did your automobile get together?"

"I guess it was put together by an auto company employee."

"Both employees were performing services. Everyone is in the service business. We are all, as individuals, performing services. You think that manufacturing is machines and administration is people."

"But manufacturing *is* machines."

"Machines certainly are part of it. How about offices? A typewriter is a machine; so is a copier, a telephone, a pencil, a dictating machine, a computer, a stamp machine, and on and on.

"The real problem comes about because the perception exists that office work, or functions like marketing and employee relations, can't be accomplished to procedures and specifications. Therefore, they have the privilege of being sloppy and wasteful if they want to be.

"Because of this mentality the price of nonconformance in these 'service' operations is twice what it is in 'manufacturing.'

"The only employees not in the service business are those who are professional blood donors; they are a resource."

## COMMENT

There *is* a difference between certain types of companies—between a hotel, for instance, and a foundry. When a customer deals with a hotel he or she meets all the employees at the bottom of the organization: doorman, bellperson, desk clerk, waiter, magazine clerk, and so forth. It is possible to travel an entire career and never meet a hotel manager.

However, when dealing with a foundry, contact is with top levels of the company. The people with suits, ties, and college educations are the ones who handle the business.

# 14

# The Saving of Emory Spellman

Emory Spellman smiled to himself as his senior staff gathered in the conference room. They didn't know what he was going to lead them into. They were undoubtedly expecting one of his detailed lectures on cutting corners and squeezing out profits. They had no idea that he wanted to make money by doing things right.

He mentally took roll as they came in:

Helen Douglas, materials. She had been a buyer before and had now been in her new job only a few weeks. Human resources (which used to be personnel) had convinced Emory that they needed a woman executive in addition to Barbara Wilson, who ran retail merchandising. Much to Emory's surprise Helen was doing a great job. Last week he thought she was moving too fast, but now he was going to get things moving more quickly.

Bjorn Anderson, production. He was always able to get the products out somehow, regardless. Resourceful was a one-word description of Bjorn, the old-timer.

Harrison Ellis, quality. Harrison prided himself on being able to identify those nonconformances that really counted and those that could be overlooked. He felt that he was protecting the company from waste through unnecessary scrap or rework. Emory could see that this was the wrong path for Consumer Consumptions. It was proving to be very expensive in the credit card areas particularly.

Bill Davis, field service. Bill was a bright and energetic modern executive. He liked to state that his people could thread a needle in the midst of a blizzard if it was necessary to keep a customer happy. Each year he had an annual awards dinner for those people he felt met the "elite" criteria of the organization.

Barbara Wilson, retail merchandising. Barbara was one of the best all-round executives Emory had ever known. She could get people to do anything and the customers loved her.

Carlton Overton, finance. Carlton was a broad-based finance person in an accounting job. He was very unhappy and about ready to quit. Emory had gotten wind of this and had planned to fire Carlton this week. However, now he had other plans for him.

"I hope it didn't make too many problems for each of you when I called this meeting," began Emory, "but it is essential that we all get together today.

"Jacob's death left us with a temporary hole in the organization that I will fill for the time being. We will all take a look at the organization together before doing anything permanent."

"Jacob was a good man. He could always be counted on to make decisions," said Bjorn. "I really miss him."

Emory nodded in agreement as did the rest of the group.

"We are going to have to make some decisions ourselves. The main problem we have as a company is quality. We discussed it briefly a few days ago. However, I have a new insight to the situation now and I feel we better get serious about quality. But there is a test I would like to have us all take. It has to do with companies that always have problems with quality."

"I saw this evaluation before, Emory," said Bill Davis. "I agreed that we usually didn't conform to all the requirements, but I have to note that field service is a profitable operation."

Carlton Overton looked up from his notebook.

"Field service is profitable only because we don't charge it the full overhead it deserves. The only real profit you make comes from the regular after-warranty service contracts."

"He's right, Bill," said Emory. "But before we get to arguing, let's go over the items one at a time. I think it will help our discussion.

"How many feel we normally deliver nonconforming products and services to our customers?"

Every hand was raised except Bjorn Anderson's. He shook his head.

"We never do things like that in production on purpose. And nothing goes anywhere unless it is covered by an official document."

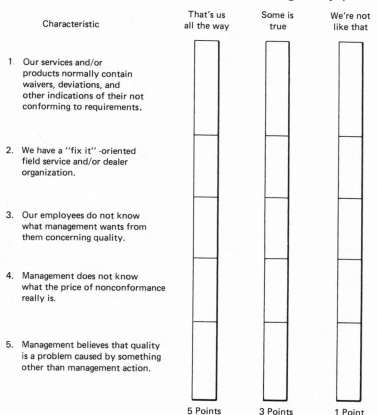

| Characteristic | That's us all the way | Some is true | We're not like that |
|---|---|---|---|
| 1. Our services and/or products normally contain waivers, deviations, and other indications of their not conforming to requirements. | | | |
| 2. We have a "fix it"-oriented field service and/or dealer organization. | | | |
| 3. Our employees do not know what management wants from them concerning quality. | | | |
| 4. Management does not know what the price of nonconformance really is. | | | |
| 5. Management believes that quality is a problem caused by something other than management action. | | | |
| | 5 Points | 3 Points | 1 Point |

Point count condition
| | | |
|---|---|---|
| 21 − 25 | Critical: | Needs intensive care immediately. |
| 16 − 20 | Guarded: | Needs life support system hookup. |
| 11 − 15 | Resting: | Needs medication and attention. |
| 6 − 10 | Healing: | Needs regular checkup. |
| 5 | Whole: | Needs counseling. |

The profile of a quality-troubled company.

"Even if it is covered by a waiver, it is still a nonconformance. If we don't put it in the advertisement, it shouldn't be part of the product," said Helen with vigor.

"Well, our service operations certainly don't always conform to any requirements I know about. The complaints we get on the credit card operations are not to be believed," commented Ellis.

"I take it that we plead guilty to number one?" asked Emory.

Reading the nods all around as agreement, Emory continued, "The company has a large 'fix it'–oriented field service/dealer organization. . . ." he read. "That certainly fits us. Now don't get defensive, Bill. If your people weren't so great we would be having a disaster right now. What this test shows is that we actually plan not to do things right and count on your folks to keep us straight with the customers."

"If you put it that way, I guess you are correct. We do feel a little put upon, particularly since the software business got so big."

"Okay, we have maxed out on the first two. Number three says that we don't have a performance standard all the employees can understand."

"I'm not so sure about that," said Harrison Ellis. "We have made it very clear that we expect quality work out of everyone."

"Fair enough. Let's take this test at its word. Each of you take a sheet of paper and write down your definition of quality, right now." Emory held up his hand to stop any comments. "Write all you want, but no talking out loud."

After a few moments he collected the papers and read the results:

" 'Quality is satisfying the customer's need.'

" 'Quality is producing something that fulfills the need of the user.'

" 'When the customer is happy that is quality.'

" 'The proper service at the proper price is what quality means.'

" 'The right amount of money spent to produce the right amount of results.'

" 'Meeting the customer's perception.'

"Aside from determining which is the right definition, it is apparent that we do not agree on what quality is or isn't. If *we* don't, then the employees obviously have no chance of knowing what quality means."

"What was your definition, Emory?" said Harrison. "Did you agree with any of them?"

Emory paused and thought about it. He couldn't expect these people to deal with the fantasy he had been through. They would have him certified and put out of harm's way.

"I have a different definition than I did the last time we met. This isn't the proper time to get involved in it in depth, but that definition is: 'Quality is conformance to requirements.' "

"But there isn't any room for considering all the circumstances if you call it that," said Bjorn.

"As I said, we can't get into it now, but we will exhaust the subject before we are through, believe me."

"The fourth item says that management doesn't know what all this nonconformance costs it," said Carlton. "That is absolutely true. I have been doing a little checking and I have a good idea what is involved. Let me ask each of you to guess what all the nonconformances cost us—as a percent of sales."

Everyone thought a moment and then gave him answers ranging from 3 percent (Bjorn) to 8 percent (Helen). Harrison thought it was about 5.5 percent.

"Well, I don't have it all complete, and there is a great deal I don't understand about it. However, I estimate that it is around 25 percent of sales."

There was a sudden intake of air all around the room. Harrison Ellis recovered first.

"I went through the examination just last year and didn't come up with anything like that. Are you sure you are listing the right things?"

"Harrison," said Carlton patiently, "field service alone costs us 5 percent of sales for the 'in-warranty' work they do."

"I put that down as a cost of doing business," said Ellis.

Emory continued the test. "The next item says that management doesn't accept 100 percent of responsibility for being the cause of the problem. Now during our last meeting I recall that we agreed exactly that it wasn't our fault. Everyone still feel that way?"

Harrison Ellis indicated that he wished to speak first.

"I have thought a lot about this, Emory. It is a cliché to say that everything is management's fault. Obviously we are responsible for whatever happens to the company. However, we don't run the public school system, we don't run the government, we don't run Japan. How can we be expected to be responsible for everything that goes wrong?"

The group all stared at Ellis. No one had heard him be this outspoken before. Bjorn Anderson was nodding in agreement.

"I think Harrison is right. We have to accept our share of the problem, but we don't make everything happen."

Barbara Wilson shook her head. "The comment was that management denies responsibility for the situation. That means that they make no effort to change their own ways; they want to change other people's ways. So when quality gets to be a problem, they send other people to school and blame institutions for causing it all."

"Are you saying that management actually causes the problems, not just lets them exist by looking the other way?" said Helen.

"I'm saying that this is what the text implies. I haven't made up my own mind yet. But let me run a little check. Bill, what is the biggest single problem you have in my area at this time?"

"That's easy. We have trucks full of no-shrink underwear that shrinks. They are being returned from stores all over the East Coast. We have spent half our time on this problem in the past two weeks. But you knew that."

Barbara nodded.

"That's right, I did know that because my telephone has been burning up because of it. Now, Ellis, how come we didn't catch this problem in our new product-testing process?"

Ellis blushed. "Well, you and Jacob called me and said that we would miss the market window if we took time to certify this product. It was coming from a supplier we never used before. In fact, it was from a country we never used before, but I let it go by based on that priority. It turned out to be a bad idea."

One more thing for poor Jacob to work on, thought Emory.

Bjorn interrupted. "What is your point, Barbara? You were within your rights in making that decision. We win some and we lose some."

"No, Bjorn, I was not within my rights. We set up that evaluation procedure to protect the customer and the company. We not only did neither, we managed to burn up a lot of money in the process. I say that here is a time where management truly reached out and caused an embarrassing, expensive, and silly problem. We created it, Jacob and I, because we were thinking in one dimension."

Harrison thought about it for a moment. "I should have refused to go along with the request. I guess I am so wound up with moving the company along that I forget that quality is my first responsibility."

"We all have that problem, Harrison," said Emory. "In the future we will help you stick to the procedures. I can think of a dozen cases along those lines."

"As long as we are doing some confessing," said Carlton, "let me say that I have just determined that 85 percent of our accounts receivable overdues were caused by action we as management took. Do you remember when we decided to tighten up on our discounts?"

Nods all around.

"Well, that changed the wording on all existing invoices. So people are merely writing us for clarification and such while holding onto the money."

"I guess we all agree that we have to correct ourselves before we can do much with the rest of the outfit. We have a great future in CC and it is my responsibility to see that the future happens," said Emory. "We are going to 'dehassle' this company, and we are going to begin with me. I want a list of all the things I have set up that bug people."

The staff looked at each other hesitantly. There always was a question with Emory of whether he was really going to change his spots or not. A few years ago they went through "sensitivity" training until it became apparent that an open society was not what the boss really wanted.

"I know you doubt my credibility and with good reason," said Emory. "But I have had an experience that has taught me a great deal about the importance of not causing problems for ourselves. I mean to take advantage of it."

"What kind of experience?" asked Helen.

Emory paused and thought about it.

"Let's just say it was a spiritual one. Suppose you suddenly realized that you might have to stand responsible for all the problems you had caused employees and customers? By that I mean if you made a decision somewhere that caused an employee to feel put down, then you could never rest until that was corrected. It would also mean you would have to soothe every dog the person kicked on the way home.

"And suppose you had to personally correct each and every non-conformance you had foisted on a customer. If those things were in your future, would you be interested in preventing them from happening?"

"You mean that you had a dream or something that every problem we cause as managers will have to be fixed by us personally before we gain any peace in the hereafter? Did you really get that message?"

Emory blushed. "I know it sounds like Charles Dickens, but I did come to that understanding. And whether that was the message or not, I did realize that I have been a very short-ranged, insensitive manager over the years. I intend to change all that."

"Well, you won't have any problems with us," said Harrison. "However, it is easy to launch a crusade for quality improvement and dehassling, but we need a lot more information if we are going to get it done properly. It doesn't take any more time to fix something that doesn't need fixing than it does to fix something that needs fixing."

They looked at him blankly.

"Anyway," said Emory, "we are going to have to get started. How should we begin? Should we put a task force together? Should we hire a consultant? What should we do first?"

Barbara spoke up. "I suggest that we get ourselves educated and gather some information while we are at it. What we think is hassle may not be, and the process of quality improvement is a lot more complex than it appears."

"A lot more people are involved than ourselves," said Bill Davis. "We only have half our direct reports here, and there are a couple of dozen other 'thought leaders' around the company. All of us have to have the same basic understanding of what this is all about. For instance, is the process of dehassling compatible with quality improvement?"

"Identical as far as I know," said Emory, "but we will find out."

"I'd like to take an active role in whatever we are going to do," said Davis.

"Okay, Bill. I'll appoint you, Harrison, and Barbara as a team of three to go find out what is available out there in the world to help us with improvement efforts. We need schooling, some consulting help, and a philosophy we can understand."

"I agree with that, Emory, but we need even more. We have to have an overall strategy. If we don't have a clear path, we will all drop out as soon as we have our first successes."

"What do you mean by a strategy? We are going to be very determined and we are going to get educated. After that we just implement—why is it different from any other management process?" asked Harrison.

"What's different is that we are setting out to deliberately change what has been a successful culture in this company. We are changing because we have to keep ahead of the times. Service alone won't cut it anymore; costs have to be reduced because price increases are just not in the cards; we have to produce products and services that look like the advertisements—every time; and we have to eliminate the things we do that hassle the employees, including us."

"That sounds like a bunch of work," said Helen.

"It really is," said Emory. "But it is all we have to do, besides run the company. We have to recognize that there is no in-between on this subject. We are either 100 percent into it or we are not. There are no various levels."

The "team of three," as they began to call themselves, did their research and made an arrangement with a firm that followed the philosophy of the Absolutes of Quality Management. The first action taken was to help the comptroller calculate the price of nonconformance to drive the stake in the ground so the company would know what its costs were at the beginning.

The second action was to bring all the thought leaders of the company—it turned out there were thirty-eight—together for a one-day corrective action session. The session was arranged on a weekend and first provided the attendees with an overview of the situation with a brief introduction to the concepts involved. Then they were divided into five groups and asked to identify the ten biggest problems of quality in the company. After their deliberations, the supporting firm took the items and compiled them into a common set of problems. The teams agreed to the summary sheet.

The ten problems listed were:

1. There is no company posture or policy on quality.
2. Management is not serious about quality.
3. There is no feedback from the field on problems.
4. Middle management does not have the authority to insist on conformance to requirements.
5. Service and administrative operations do not have written procedures.
6. Customers do not order from catalogs; there are too many specials.
7. Morale is very poor because people don't trust those of us in management; they feel we pick on them.
8. We need new services to offer.

9. Not enough money is being spent on research and development.
10. Corrective action is very hard to get; no one looks past this week.

The items were discussed, and a workshop was conducted on the cost of quality so the managers would understand what it was. Then the groups were re-formed and asked to offer some suggestions for action that could be taken on the problem items. The primary suggestions were:

1. We need to have the tools to turn quality around.
2. Let's find a way to ask the people what bugs them and then eliminate all we can of that.
3. The field service reports need to be analyzed and then shared with us.
4. We need an overall company quality improvement program.

The team of three discussed the status of their efforts with the group, and the general feeling was that this was the correct track to take. The group identified a total of fifty-six managers and other key people who had to be educated immediately. It was decided to appoint a quality improvement team after the executive education was completed.

Each of the selected executives was sent off to an executive quality management session for two and one-half days. The members of the quality improvement team were sent to the management college to learn how to run the process. At the same time, the in-house education system to be given to all employees was obtained and begun. The support firm supplied the material and trained the instructors.

Within three months, the assigned company managerial people had all been to school and all shared a common language of quality. Everyone was talking about "meeting requirements" on PONC and other communication specifics. Field service reports were dramatically specific.

Suddenly the veils were falling and the real problems of the business were revealing themselves. As the quality improvement education effort hit each department, people began to realize that there was no need to put up with the hassle that had been going on. They realized that it was all right suddenly to admit to a problem and begin to do something about it. They understood that management actually wanted them to bring things to the surface and eliminate the cause.

Emory was delighted with the pace of progress. The enthusiasm was genuine.

"We are off to a good start, Carlton," he told the comptroller. "This isn't motivation; it's an attitude change. We are beginning to turn the place around. It sort of feeds on itself."

"The figures show it too, Emory," said Carlton. "We are seeing a lot less rework already. And the salespeople have even asked for a special class on how to write orders properly."

"I am anxious to see how the supplier quality meetings will turn out. If we could just get our suppliers as interested as we are, we would be able to make a big jump ahead."

Helen walked up just as this comment was made. "I want you to know that they are very enthusiastic just because we are having the meeting. The results will be excellent.

"However, I am learning that some of our biggest suppliers are right here inside the company. Do you realize that we process several thousand internal documents every day and that almost all of them have an error or two? *There* is a fertile field. And the credit card problems that we cause ourselves? Wow!"

"I never thought of it that way," said Emory. "You think there is a lot of internal error in paperwork?"

"Let me walk you through something that we have picked up over in fulfillment. It shows the problem we have in the whole company.

"As you know, we receive 5,000 purchase orders every day from the field operations. We process these through the computer, pull the individual merchandise out of stock, pack it into the boxes, and mail the boxes to the salespeople. The salespeople take it to the customers."

"That's the way we set it up years ago," said Emory. "It has worked very well."

"Yes," said Helen, "it does work well. We have very few returns from the salespeople and put-through time is pretty standard at two days. One would get the impression that this is a model of efficiency."

"By now," said Emory, "I recognize that there are very few models of efficiency. You remind me of a ghost I met. I have the feeling that I am going to learn all kinds of terrible things about one of my favorite operations."

"You met a ghost?" asked Helen.

"Just a figure of speech. I was daydreaming one day and I felt that Jacob came back to tell me that if I didn't get my act together, I would suffer through eternity."

"Is that what made you start this quality improvement process?"

"It helped. I guess it took that thought process to make me realize that change was necessary. Anyway, tell me more about this 'ghost of quality present.'"

Helen smiled. "Well, we have 258 people in this operation, from those who open the letters to those who shove the boxes on a truck."

"Yes, we've kept it small over the years in order to make it cost-efficient."

"Well, I did a breakdown on what those people are doing. I think you will find it interesting," said Helen. [She showed Emory the table on page 137.] "And that is just one operation."

Emory was stunned. "But we have always thought of this as a well-run place."

"It is well run. But it's run according to the 'be resourceful and keep it going' school of thought. We don't work on prevention; we work on fixing. We have never trained the sales personnel on how to fill out forms. All they get is a letter.

"And they have learned that someone will call them if there is a problem, so they don't take it very seriously. We have a request for a whole new computer system just to take care of that."

"So what specifically is going to happen in this situation? I can see several corrective actions that need to happen, but they would just be patches. How do we change the whole place around?"

Helen pulled out a schedule of events.

"Since this is a fairly isolated operation, we decided to treat it as a project of its own. It will benefit from the overall company commitment, but we will be able to go after the improvement process in a local way.

"Here are the steps we are going to take:

"1. Determine the price of nonconformance.
"2. Educate the senior management.
"3. Select the quality improvement team and send them to receive education.

| | People |
|---|:---:|
| Opening mail, sending checks to accounting and POs to computer entry. | 5 |
| Checking POs to see whether the data on them is correct for computer entry. | 5 |
| Calling salespeople to clarify data; marking order blanks if wrong pencil was used; doing general rework. | 12 |
| Entering POs into the computer so orders can be sent to the factory, billing to accounting, and receipts to salespeople. | 2 |
| Accounting, billing, and other financial for planned work. | 31 |
| Financial to verify bills with salespeople. | 6 |
| Internal mail and services personnel. | 29 |
| Order entry—making certain that the order being sent to the warehouse is accurate. | 11 |
| Warehouse screening, setting up the orders to be processed. | 15 |
| Warehouse computer operations. | 9 |
| Warehouse inventory operations and order fulfillment. | 69 |
| Shipping operations. | 23 |
| Security. | 9 |
| General operations (personnel, etc.). | 32 |
| | 258 |
| Personnel involved in checking, reworking, or special handling, all classified as price of nonconformance. | −96 |
| | 162 |
| Typical wages plus benefits and other overhead | $   36,000 |

$$\$36,000 \times 96 = \$3,456,000$$

"4. Lay out the improvement strategy.

"5. Educate the other personnel internally, using the employee education system to bring them all up to date.

"6. While this is going on, we can establish the first six steps of the quality improvement process. This will lead us in choosing the building blocks of the necessary system analysis. The corrective action will then be to identify and eliminate the problems forever."

Emory nodded in approval. "Let me see whether I have this straight. First you are going to establish the environment for caus-

ing improvement, by educating people and at the same time formalizing the management commitment, the measurement, and
the awareness."

Helen nodded.

"Then you are going to examine the whole business of processing the purchase orders. This examination will be successful because the people involved will all understand quality and hassle the
same way and will all have an understanding of their personal role
in causing quality improvement.

"Doing it this way will succeed because everyone will be interested in causing quality improvement instead of fighting it."

"Emory," smiled Helen, "it sounds as though you wrote the
book. You summed it up exactly."

"I'll be watching to see how it all comes about. What can I do to
help?"

Helen paused. She looked at the table top, then at Emory's coat
lapel, then at the window, and back to the table top.

"I have the feeling that you would like to tell me what I can do to
help but are embarrassed. I must have done something that I
shouldn't have. Let me take a few guesses."

Helen nodded and produced a tentative smile.

Emory sat back and gave the matter some real thought. He had
been busy going around to each area talking about quality. In this
process, he was always careful to be very positive about quality and
his personal commitment to it. So that couldn't be it. He had quit
talking about the "economics of quality" and the "customer's perception," so it had to be something else, something specific.

He turned to face his visitor.

"Could we be talking about the answer I gave to the comptroller
when he asked me whether we should continue to allow so much
for callbacks by the service people? I told him that I thought if a
service person didn't have a callback now and then, he or she was
working too slowly. I guess that wasn't the right answer?"

Helen sighed. "Now that you bring it up, it did have the result of
making them think callbacks are a normal business expense. They
are planning a program to reduce callbacks rather than eliminate
them."

"Boy, words are important. I'm really sorry about that. I will see
that the confusion is straightened out."

"That will be a big help," said Helen. "We are having special training of the customer service people on this subject. They need to get their attitude lined up with the thought that they are a profit center. Bill is working hard on them."

"I'll be anxious to see how fulfillment turns out."

"We'll keep you informed on a day-to-day basis."

"Don't waste time on me; every few weeks will be fine."

## FULFILLMENT QUALITY IMPROVEMENT REPORT: NUMBER ONE

### Education

All executive and management college sessions are complete. A total of twenty-four participated.

Remaining personnel are entering the quality education system sessions. The instructors were trained last month.

Problem: Pulling together examples so the QES instructors can relate real life to the students. Our resource company has been very helpful in this. One of its counselors spent two days with us and we have enough now.

### Improvements

1. Getting salespeople to fill out the purchase orders exactly correctly is the number one problem. At least four rework operations are caused by this problem. Twenty-four jobs exist for the sole purpose of getting that information accurate.

   *Action taken:* Marketing has agreed to simplify the form by placing the least-used two-thirds of the items on a separate form that does not change every month.

   Each salesperson will receive a special booklet explaining in a graphic manner exactly how to fill out the form and offering a credit for sending in "Zero Defect" forms. Credits will accumulate.

   Each communication to salespeople will contain a small No. 2 pencil. The largest single problem we have in this area is caused by salespeople not using that type of pencil. The computer can't read anything else.

   Long-term salespeople will have the privilege of transmitting directly to the computer.

   *Results·* To date the purchase order error rate has dropped from a traditional 23 percent to 9 percent. The major impact has yet to be felt. Response from the salespeople is very positive. We were able to transfer fourteen people to the credit card division, which had been advertising for exactly the kind of people we were using for rework.

2. We found that there was a regular 20 percent overtime requirement in packaging operations due to the absence of box lids late in the afternoon. Employee error-cause removal ideas pointed out that the box lids were diverted to provide inventory for start-up the next morning. There was no good reason for this that anyone could find, so the situation was changed. Overtime is no longer necessary.

In order to identify the opportunities for improvement and for cost eliminating, we are beginning a BAD program next month. BAD means Buck A Day (copyright © by Industrial Motivation Inc., New York). It should provide us with a way of getting all employees to let us know what our problems are.

## FULFILLMENT QUALITY IMPROVEMENT REPORT: NUMBER TWO

### Education

All personnel have gone through QES and we are beginning to run it for the credit card division. Personnel have been to instructor school and have examples, but we find that working together takes the load off. The response of our students has been dramatic.

### Improvements

1. As reported by the finance office, PONC has been reduced by 37 percent in these first six months. Purchase order defect rate is now 2 percent with the biggest problem still being the nonuse of the No. 2 pencil. Research has accepted the task of eliminating that requirement.
2. BAD month produced 312 suggestions from 246 employees. All are excellent. Everyone really liked the thought of trying to save a dollar a day in each job. We estimate that these ideas will produce better than a half-million dollars in direct savings this year.
3. We have so many specific items that we are attaching a separate page.

During the first year of the quality improvement process, CC reduced its price of nonconformance from its calculated level of 26 percent to 16 percent, comparing "apples to apples." The hassle level of the company dropped by more than half, according to Emory's unscientific index. Morale was up, customer complaints were down by 75 percent, and the field operations were getting interested in maintenance instead of failure.

Emory's personal calibration of the credit card division related to how many irate telephone calls or letters he received. Those had begun to disappear.

Overall he felt terrific—everything was going well. The first QIT was about to hand over control to a new group, and the Zero Defects Day he had reluctantly agreed to have turned out to be a smash. The senior management team had quit complaining about quality and about each other. It was like a dream.

Emory sat up with a start. Was all this another one of Jacob's visions?

No, he satisfied himself that it was real. But the thought pulled him back to reality.

There had been a great deal of improvement and things were going the right way. But the surface had only been scratched, the tip of the iceberg only sighted. They were still in deep trouble.

"Schedule a staff meeting on quality for tomorrow morning," he shouted into the phone. "We have to get serious about quality."

Emory wasn't going to spend eternity in that warehouse.

# 15

# The Dehassling of Lightblue

Harrison Wilson stepped briskly into the boss's office. The expression of concern on his face had long since been replaced by one of puzzlement. He nodded in greeting to administrative assistant Carol Bennington and walked over to stand beside a smiling David Rocque.

"David, I can't tell you how embarrassing all this is to me. I really hope it can all get clarified before the book comes out."

"Don't take it all so seriously, Harrison. These things happen. What the man wrote about our planning system is just his opinion. People will read it with interest and that will be it. The things that deserve to live keep on living. There isn't much you can do about it anyway."

Harrison moved across the room. "That's just the point, David. Does it all deserve to live? It isn't what the author said about our planning system that disturbed me; it's what our executives told him about it that gives me a problem. I have been studying that chapter ever since he sent it in for approval. We have a big problem."

"But if you don't want all that stuff said about how the system has no earthly use, then just disapprove of the material. That will be the end of it."

"There's nothing wrong with the material. It says that the planning system I have so carefully conceived and implemented is a useless hassle. Unfortunately, most of our executives agree with him, and even worse, so do I. Until now I never looked at it all in complete terms. Frankly, what we are getting out of it is not worth what we are putting into it. We—I—have created a cure for which there is no disease."

Harrison sat down in a chair and slumped dejectedly.

Rocque patted his shoulder. "Don't be so hard on yourself. We would not have been able to grow as we have if our executives had not learned how to accumulate and analyze information the way you taught them. The fact that the system became a little cumbersome does not detract from its usefulness. Every method has to keep up with its customers. Just go back and take a look at the system. 'Dehassle' it, don't destroy it."

Harrison looked up gratefully. "I'm glad you aren't disturbed. This could make us look a little thick when it comes out."

"Only if we have not recognized the need for change and improvement. I have felt for some time that we were outgrowing our planning system, but I figured that it was still a good investment and was serving a primarily educational function. I still think it is the only place the whole company comes together."

He moved over and sat beside the younger man. "A company is a growing, living entity with its own personality. It only gets into trouble when the management doesn't recognize change and lets the company outgrow its clothes, so to speak.

"We have updated our hiring practices based on society today; we have modified our retirement programs to take advantage of IRAs and other government initiatives; we have sold off several product lines that were becoming obsolete (and we wouldn't have recognized that if it hadn't been for the completeness of the planning system); and we have changed our whole internal communication system based on the availability of personal computers and the cable systems.

"So we have outgrown and rebuilt a lot of systems. We do that on a routine basis. Planning is just one part of our management practice; it needs to serve us rather than the other way around."

Harrison was beginning to feel a little better. "Actually, we have been blocking out a system of including the planning material as part of the electronic mail system. We could eliminate 70 percent of the detail meetings by doing that."

"That would be a good beginning."

"But I think that before we do anything I am going to bring some of the key senior people together and ask them what they feel they really need in order to plan properly and, in fact, to run the corporation properly. We still don't know exactly what it takes to bring it all together." Harrison stood up and moved to the window.

"I think we can identify the majority of category items for the routine planning items. The ones that concern me relate to the identification of the future. We almost missed the evolvement of these new life insurance policies because we weren't looking at insurance as a short-term investment."

Harrison nodded. "And we didn't pick up on the need for the large-scale integration in semiconductors early enough. We never have caught up in that area. It's hard to watch so many markets at once."

"That is the point," said Rocque. "It is hard to watch all those markets at once, and it is hard to know what is going on in an industry at any given moment. But the people who do that exclusively seem to be fairly close to the truth on a consistent basis. No one knows for sure, but those who live with the subject all the time have a solid input. However, I think that input needs to be viewed by someone who is looking at the overall business world."

"You mean if someone in the hotel area determines, for instance, that more businesspeople will be taking their spouses on trips, someone outside the industry might point out that two-career families could cause a trend the other way?"

Rocque nodded. "Right. Without putting any constraints on executives, we need to build a broad information system that will let us see all sides of trends or all the effects of planned actions. After all, there are no islands left today."

Wilson had his old energy back. "I think we could restructure the planning concept and the support operation so it would require about a third of the effort and produce more practical results. But first I'll bring together that group I was talking about. We will never let planning of any kind isolate itself again. And we will certainly never let it get to the point where it actually causes hassle. Imagine, something set up to eliminate problems being one of the major causes of problems.

"We will be able to reduce staff; that's a sure thing. I'll be willing to walk the plank myself if that will help."

Rocque was startled. "There's no need for that. You are the best equipped to get us on the correct path. We will change a dozen more times before you retire in twenty years. So don't be sensitive—we only learn from what happens to us."

In preparation for the key executive meeting, Wilson laid out a communication format. He divided the planning mission into three areas:

1. What do you need to know in order to run your operation?
2. What is needed to provide a view of the corporation overall?
3. What sort of "nice to know" information should we acquire?

Much to his surprise, Wilson found that the line executives were more concerned about getting a strong planning concept than the functional and staff management.

"We need to know where we are going, and where we have been," said Deborah. "However, we only need to know in broad terms; it's not necessary to have each little thing laid out."

"What little things would you avoid, Deborah?" asked Howard Gibbons. "One person's 'little thing' is another's 'big deal.' What about monthly sales forecasts?"

"Monthly reports would be proper for most things; it's weekly and daily reports that drive us bananas. One of the most time-consuming items is inventory. I would like us to concentrate on eliminating inventory altogether rather than finding better ways to keep track of it."

It was Walter Thomas who clarified the situation for Harrison. "I think there might be a basic misunderstanding all around on this subject. It isn't the collection of data that causes all the dissension. I have never met any operating executive who felt there was enough information available.

"And it isn't the assembling and presenting of all that material. It is the damned hassle."

Harrison looked blank. "We are going to simplify all the forms and procedures."

Walter smiled patiently. "Personal hassle, not forms."

Deborah Greese interrupted. "It is the business of having to go to all those meetings where we are only spear carriers. It is having those staff people keep track of the program of every characteristic and then take up our time and people asking why we ate the carrots before we ate the peas."

Carl Watson agreed. "As a functional director I have to sit in while each and every subject is discussed. At least Deborah only

has to get involved with the insurance and finance discussions. I'm
in for it all."

"You have to be in the meetings," said Deborah, "and I sym-
pathize with you. But I'm sitting out in the lounge trying to run my
operation from an extension telephone."

Harrison winced. "Well, at least we all get to see each other at
these meetings. That has to count for something."

"And it does, it does," said Gibbons. "But do they have to be so
frequent and so painful? Did you look at the chart that was made
up on us? (*See* Chapter 3.) According to that, we only have one
day a year to do our work."

"But we have to develop the plan. I realize that it takes much
time and that there is overlap. But how can we agree on what we
are going to do if we don't ever meet?"

"I think the meetings will take care of themselves if we get the
basic concept right. I think the problem is not that complex. We
have to agree that the purpose of the planning activity is to help the
operating and functional departments do their job."

"I think that's obvious," bristled Harrison. "What else is it for?"

"You could get an argument about that, Harrison. I don't know
of anyone who thinks that it's for any reason other than to help
corporate keep track of the troops."

"No wonder people keep talking about hassle. If the entire process is perceived in that manner, then we have really done a lousy
job of presenting it. You make it sound like a punishment from
God."

"I doubt that anyone has laid this burden at the feet of the
Almighty," said Carl. "But the fact is that punishment is the correct word. We have somehow turned an essential tool for progress
into a scene from *The Count of Monte Cristo*."

"So what should we do?" asked Wilson. "We need to collect and
distribute basic data; we need to project the financial plan for the
company; we need to schedule our work; and we need to agree on
the resource allocation. We have to know where the company is
going and make certain that it is getting there."

"All that is true. That is what our planning system should be
designed to accomplish. Set it up that way and everyone will be
happy."

"But it *is* that way," said Harrison. "All those things are what we
have laid on now."

"What is laid on now is 20 percent of those things and 80 percent beating up on the executive corps. The entire process is one way."

"That was never the intention. The whole idea was to find the most efficient way of getting the best plan and ensuring that it was met." Harrison reflected a moment. "Come to think of it," he said, "that is probably what every totalitarian government says."

"You got it," said Deborah.

"So what we are saying is that we have no disagreement about the need for a corporate plan, and we don't even really have a problem about needing functional specialists to help meet the plan."

Nods by all.

"The problem arises in the system that has developed, in which the entire purpose of the corporation is to create a plan and publish it.

"When that is the purpose of the company, then executive time and energy is wholly dedicated to that work. All niceties, like being able to spend time doing the things you think are important, become secondary. We have become *plan worshipers*.

"Well, we have the chance to start over. We will design a system and do it together. That helps the operating and functional management get their job done and at the same time lets corporate management agree with their direction and have enough information to fulfill their obligations."

Harrison looked around to see how this had been taken.

"That's the right course for us, Harrison," said Deborah. "There is no reason that work should ever be unpleasant. Here is my list of what we need to run insurance and finance. Now that the philosophical discussion is over, I suggest we all get on this planning of the plan. I have the feeling that it won't take long at all."

# 16
# Some Success Stories

Some years ago, I worked with two U.S. plants of the same corporation to help them be the pioneers of quality improvement. One plant was a telephone equipment manufacturer, and one made pumps. Both had dedicated general managers and both followed instructions on how to cause quality improvement.

In the course of the first nine months, they reduced manufacturing error rates by more than half, they reduced their cost of quality by 20 percent, and they completely eliminated customer complaints. By any measure that could be selected, they were successful.

Since the process was installed by existing personnel, the only expense involved was for some communication material. Both facilities became really profitable for the first time and were recognized for their achievement by senior management.

So we figured that this was our opportunity to reach the rest of the North American operations. We had begun the same strategy in Europe and were attaining the same results.

Therefore, we arranged a quality seminar for all the North American division presidents, plant managers, and group people. At this seminar the successful executives explained what they had done, showed the results, and answered questions. The staff stayed in the background to ensure the credibility of the project.

The result was that some more people were convinced that a quality improvement process might help them out. Most, however, felt that it was all right for those folks who had easier jobs but it wouldn't work in their locations, which were very difficult. A few stepped out and did well.

We all know that it is not enough for something to be worthwhile. We have to be personally convinced that we cannot live without it.

So we had to deal with each and every unit individually, showing them the error of their ways by pointing out the costs of being bad and the virtues of improvement. After a few years, the evidence became overwhelming and the units began to demand support.

There is some value in talking about what others have done, so I am going to describe some of the successes we have seen in the past three years. The information was gathered by the various PCA counselors and supplied by the client companies. For obvious reasons, I am not going to list company names, but the stories are all documented and accurate.

I won't include "communications" items, since every single company reported them. What they all said was that communications between different company entities, like corporate and location, or staff and line, had improved dramatically. That is a typical reaction.

## COST OF QUALITY

### Computer Manufacturer

The pilot site, producing peripheral equipment, reported comptroller-calculated cost-of-quality reduction of $241 million after twenty-two months of the quality improvement process.

Production had increased 48 percent and the field service population, as one example, had shrunk.

### Semiconductor Manufacturer

A total of $35.5 million in manufacturing costs was eliminated during a two-year period. In addition there was a $3 million reduction in material handling costs.

Administrative costs were estimated to have reduced a great deal, but since they were not documented, they are not listed. The corporate secretarial quality team, for instance, was able to reduce the total monthly cost of telephone service by 42.9 percent. (They did not want to share their telephone billing number.)

A main source of cost elimination was one of the five wafer lines, which contributed $250,000 a week due to an intensive effort to follow the existing process. The other lines now believe and are following the same thought pattern.

## Sweeper Manufacturer

This $90-million sales company has reduced its cost of quality each year. Beginning with 16 percent in 1979, the cost of quality dropped to 11.5 percent in 1981 and 10 percent in 1982. The goal for 1988 is 2.5 percent.

## Communications Company

This $14-million sales company reduced COQ from 29 percent to 20 percent in the first year.

## Multiproduct Small Company

The COQ analysis showed that one product produced 11 percent of sales and 60 percent of warranty expenses. A quality engineering evaluation revealed that the company just did not have the technical skills to deal with this product. The firm sold it and improved profitability.

## Oil Refineries

This big-name energy company has four refineries in the U.S. It has reduced its COQ by a total of $50 million over a two-year period. All this was due to eliminating wastes such as mixing in the wrong additives and doing maintenance over, and to eliminating excessive hydrogen spillover and generally concentrating on process discipline.

One turnover project planned and run with Zero Defects was finished early with a resulting profit improvement of $2 million.

## Textile Manufacturer

The cost of quality companywide was reduced 22 percent in two years. The interesting point is that only half the facilities were involved at that time.

## Another Textile Manufacturer

In the pilot operation, the COQ was reduced 1.6 percent in the first six months. That amounted to a $700,000 savings.

## PAPERWORK

## Sweeper Manufacturer

During 1981, invoice errors caused by the billing department were reduced by 38 percent.

## Semiconductors

The payroll department had twenty-one people and a lot of overtime. Each week there would be a long line of employees waiting to have their checks corrected. The primary cause of this was that the supervisors could never get the cards in on time, and the payroll clerks couldn't determine who was late or absent or worked overtime.

By assigning the same people's timecards to the same payroll clerk, and by setting a cut-off time for getting cards in, the firm eliminated the problem. There was no need for a chart to show the reduction. It was only necessary to see that the line had disappeared on payday. Seven people in the payroll department (of the twenty-one) were able to move on to better jobs in the company. Errors dropped from a consistent 20 percent to less than 1 percent.

In another area, the process control department began to count the number of mistakes on tags and log sheets that accompanied wafers. Over 900 errors were generated in that department, mainly due to misunderstanding of the paperwork forms and requirements. A clarification of the requirements, and some training classes, dropped the errors to below twenty a week. This happened almost immediately.

## Computer Manufacturer

By placing reports on the computer and letting people call them up when information was needed, this company reduced paper cost by $154,000.

By concentrating on getting bills paid on time, a two-person team in the finance department reduced the lost discounts to $250 from $43,000.

## DATA ENTRY ERRORS

### Chemical Company

In three weeks keypunch order entry errors were reduced from 2,300 a week to 1,300.

### Energy Company

From June to December, the entry errors decreased by 41 percent.

### Power Company

Purchasing change notices were reduced by 70 percent through enforcing requirements for complete information on requests.

### Automobile Supplier

Editing of sales orders has been discontinued, since errors have disappeared in the last three months.

## DEFECT REDUCTIONS

### Computer Manufacturer

"Off specs" were reduced from 43 percent to 3 percent. ("Off specs" is a formal request for a deviation from requirements. It had almost become routine.)

Software defects per 1,000 lines of code were reduced from sixteen to four. (Recently one team produced 20,000 lines with ZD— ahead of schedule.)

Field kits are sent out with 98 percent content reliability now; it had been considered normal to have 60 percent nonconforming.

This company had been suffering for a long time from internal printed circuit board failures and high product mortality in the field. It was convinced that the problem originated with its internal supplier of semiconductors. When the quality improvement process began, a team was assigned to the problem to evaluate it from scratch. The main problem was determined to be caused by an electrostatic discharge during the assembly process. Management had been so convinced it knew the answer, it never looked beyond its own opinion. The problem disappeared once the preventive corrective action was taken.

## Sweeper Manufacturer

Hydraulic leaks were reduced to 0.002 per machine from 2.4 per machine. Much of this was accomplished by reducing the number of fitting suppliers from twenty-eight to two, and by doing some redesign.

## Semiconductor Manufacturer

In January, a customer was rejecting 7 percent of the components received and was ready to cancel. By May the rejection rate was down to 1 percent, and in June the customer reported zero rejections. One work group processed 76,948 wafers without an error.

Scrap due to scratched wafers has dropped from 4 percent to 0.6 percent and is still being reduced.

Field failure returns of MD boards are down to 0.4 percent from 8 percent. Overall the defect rate internally has been reduced from 1.5 percent to 0.2 percent.

## FIELD SERVICE

### Appliance Manufacturer

The field service call rate for a product developed under the QIP is fourteen per 100, the lowest ever attained on a new product.

## Test Equipment Manufacturer

Twenty-three percent of printed circuit boards used to come back for one reason or another. Now less than 2 percent do.

## Semiconductor Manufacturer

The warranty on boards has been extended to a full year from 90 days due to the improvement in quality.

The field orders received went thirteen weeks with Zero Defects.

## Refinery

Service-after-service measurement revealed that seven cases a month required repeating. Management attention, training, and measurement reduced this to none in three months.

## CONCLUSION

Everything improves when a company becomes genuinely involved in the quality improvement process. The lowering of the hassle level is one of the most pleasant side-effects of the effort. Suddenly voices drop, discussions are more orderly, and the problems of quality are resolved without emotion. This is the result of having a common language of quality and a common comprehension of the goals of the organization.

The other area that shows obvious improvement is the administrative. Actions become smoother, paperwork is less cumbersome. Long-standing confusion over procedures and processes is ironed out. The cynical view that paper is just a burden to be borne begins to erode.

The successes listed above provide a guide for the type of reporting that is necessary. People need reassurance, continually, and there is no better way to accomplish that than talking about the successes that have happened.

If you don't believe this helps, just ask anyone what automobile stands for quality in his or her mind. Almost everyone will name Rolls Royce. Then ask whether the person has ever owned one, driven one, or even ridden in one. If they have not, then ask how they know that it represents quality.

Rolls Royce told them.

# 17
# Mixing the Vaccination Serum

Every ingredient involved in setting up a permanent system to eliminate hassle and to cause quality improvement requires special attention. Now that we have taken an overview of determination, education, and implementation, it is necessary to look at some of these ingredients in detail and state specifically what has to happen. All these steps were listed in Chapter 2.

## INTEGRITY A

The chief executive officer is dedicated to having the customer receive what was promised, believes that the company will prosper only when all employees feel the same way, and is determined that neither customer nor employees will be hassled.

## Determination

The CEO must continually communicate with the customers and the employees of the company to assure them of personal determination on this subject. This requires making the CEO speech mentioned before and constantly remaining physically accessible to all the people. It is important that the CEO remember during business meetings to emphasize that quality is first among equals, to insist on bringing up the downside of any particular problems that may have occurred through nonconformance, and to continually remind the senior executives that they are not to set up things that hassle the employees. All of these sound rather obvious until you remember that it only takes one phrase, such as "Quality is important but don't forget we still have to sell things," to turn the successful process back into the same old thing.

## Education
The CEO needs personal awareness to understand the particular role assigned to this job. This means executive education and enough exposure to the content of the employee education system to be able to talk about it knowledgeably. It is also the CEO's responsibility to see that the board of directors comprehends what the quality improvement process is all about. That should include participation in the education effort.

## Implementation
The CEO needs to see that the corporate policy on quality is issued, is understood, and is communicated to everyone. It is advisable for the CEO to drop in on the quality improvement team regularly and to make certain that they are not getting bogged down in some sort of useless trivia.

### INTEGRITY B

Chief operating officer believes that management performance is a complete function requiring that quality be "first among equals"—schedule and cost.

## Determination
The COO in many companies represents the voice of practicality. It is assumed that the CEO will come out in favor of long-range, broad-based, worthwhile objectives. The COO, however, has to produce sales, products, or services and keep the whole operation moving. So employees look to the COO for real values.

## Education
The COO needs the same education as the CEO but must understand the cost of quality and supplier quality management in

depth. Both these are areas that will be very fruitful in operating the company.

## Implementation

The COO must make certain to let operating entities know that quality improvement is not optional. Individual approaches to quality improvement should be encouraged, and certainly no one has to follow a specific cookbook. However, quality improvement has to be taken seriously and has to be a management process or the COO will never be able to achieve the goals assigned.

## INTEGRITY C

The senior executives, who report to those in A and B, take requirements so seriously that they can't stand deviations.

## Determination

The phrase "take requirements so seriously" means that the executives who run marketing, finance, sales, manufacturing, engineering, regional offices, legal, public relations, quality, purchasing, information services, and all the other functions have to recognize that if they are going to have quality improvement in their areas, everyone has to agree on what is going to be done and work to that end. This commitment has to be shown by these senior executives on a daily basis or the whole process will be ineffective.

## Education

Senior executives need to go to executive education and to workshops on their specific functions. The executive in charge of quality has to tie in all the measurements and reporting with what is supposed to happen in every other function.

## Implementation

Each and every function needs to have its own look at the quality improvement process as it applies to that function. Certainly someone from that function is participating in the overall company process, and that is as it should be. However, each function has its own needs. Marketing, for instance, needs to review its own operations in terms of the Absolutes of Quality Management and the techniques of quality implementation. Communication inside that function needs to be examined to see whether problems are really being identified, brought to the surface, and solved. So each function should see whether it is indeed doing its part to make itself hassle-free.

### INTEGRITY D

> The managers who work for the senior executives know that the future rests with their ability to get things done through people—right the first time.

### Determination

This level of management is where most of the difficulties arise, because these managers are the ones who have to do the actual work. They need to be continually reassured of the commitment that resides in the Integrity A, B, and C entities. Then these managers need to transmit that credibility to the rest of the employees. The determination they show is what will convince people that the company is really serious about the problem.

### Education

Managers need more working-level information about the process than the senior executives, so they need to attend the management education, which usually takes about a week and then has follow-up. They also need to participate in the employee education system as instructors, or at least as participants in the discussion sessions. Their visibility is very important. These managers would also be exposed to specific workshops on subjects like supplier quality management, inventory, or other areas that would help them do their job better.

### Implementation

Many managers of this level find themselves on quality improvement teams being responsible for helping the process to happen.

However, for the most part they will need to be the walking, talking, living example of the quality improvement process.

### INTEGRITY E

The professional employees know that the accuracy and completeness of their work determines the effectiveness of the entire work force.

## Determination

The creative force of any organization lies in its professionals. Whether it is a law office or a computer company, there is a great reliance on specific trained professionals to create the material that the company sells. The professionals identify the requirements necessary to create the material and use their professional expertise to make these requirements clear and understandable to everyone.

## Education

Professionals (except those being assigned to quality improvement teams) usually only need to go to the executive education and to the workshops specifically set up for that function. Paraprofessionals and all other employees in the professional departments need to participate in the employee education system so they can help with what is going on.

## Implementation

One of the things most valuable to the professional is feedback on the service or product from the interim or end customers. Information passed through the professionals lets them make their requirements even clearer and less ambiguous. Professional organizations must recognize that they are the thought leaders in the company, and that if they don't take quality seriously, the rest of the people won't.

### INTEGRITY F

The employees as a whole recognize that their individual commitment to the integrity of requirements is what makes the company sound.

## Determination

When all the employees are determined to conform exactly to the requirements and to offer feedback when the requirements are inadequate or impossible to achieve, then hassles will begin to die out and quality improvement will become a fact of life.

## Education

All employees except those mentioned above need to be involved with the employee education system. Each one of them needs to be exposed to a complete understanding of what quality means and how each can act to implement it. In addition, new employees coming aboard need to go through the Quality Education System (QES).

## Implementation

Quality improvement has to become a daily part of the work force's life, whether people are involved in creating insurance policies or in pouring molten steel in a foundry.

## SYSTEMS

### SYSTEM A

The quality management function is dedicated to measuring conformance to requirements and reporting any differences accurately.

## Determination

The entire professional quality function has to become committed to helping cause conformance to requirements, rather than to carefully identifying nonconformances and keeping neat records of them. In order to do this, the quality group must establish measurement capabilities at all levels of the process, whether they or someone else actually performs them. These measurements let management know where the areas of nonconformance exist. The quality function then leads the corrective action charge and insists that the quality improvement process be carried out.

## Education

The senior quality management personnel should go to the management education, and all others must attend the employee education system classes. In addition, professional education is required to ensure that the most efficient systems of measurement and reporting are utilized.

## Implementation

The quality function must become a well-disciplined and organized operation that carries on the spirit of the quality improvement process without becoming a disciplinary force. It should be oriented toward the role of prevention in all things. This means that quality must not limit itself to the operations or manufacturing activities but also must assist other functions, such as those involved in the Integrity C, D, and E systems, to help define their own internal quality systems.

### SYSTEM B

The Quality Education System (QES) ensures that all employees of the company have a common language of quality and understand their personal role in causing quality to be routine.

## Determination

Every employee of the company, without exception, must have a complete education in the understanding of quality and what it means to him or her and to the company.

## Education

The system as developed must be performed intact. Shortcuts always cause problems.

## Implementation

Usually one class a week is about right. This keeps interest up during the period of the course and lets the employees understand that it is a serious business.

## SYSTEM C

The financial method of measuring nonconformance and conformance costs is used to evaluate processes.

### Determination

Evaluating quality according to financial measurements, such as pointed out under the cost of quality, brings it into a very professional and front-level management position. This guarantees that everyone will take quality seriously. As companies become more used to this method of evaluation, they will learn how to apply it to individual processes, of paperwork as well as of production.

### Education

All operating management people certainly must understand how the cost of quality is put together and how it applies to their area.

### Implementation

The comptroller function should always lead any financial measurement system.

## SYSTEM D

The use of the company's services or product by customers is measured and reported in a manner that causes corrective action to occur.

### Determination

Many companies do not have the slightest idea what happens to their product or their service once it leaves their hands. This seems ridiculous, but there are in fact almost no fail-safe customer-usage or field-failure reporting systems. The proper system causes corrective action by showing management where problems need to be eliminated and where previous corrective action has not been effective.

## Education

Field service management needs executive education; field service personnel need the employee education system and special education in communication of problems.

## Implementation

A simple form that can be understood by everyone and used with little effort must be constructed. If the information can be input directly to the computer system, so much the better. However, the reason most present reporting systems fail is that people learn very quickly that the information is not taken seriously and that nothing much happens about it.

### SYSTEM E

The companywide emphasis on defect prevention serves as a base for continual review and planning that utilizes current and past experience to keep the past from repeating itself.

## Determination

Problem elimination needs to be examined as a part of the formal operation of the organization. These sessions must be scheduled and proper minutes maintained. The type of problems that evade solution must be identified and moved into a task-team environment. No problem must be allowed to exist and to recur.

## Education

Special short courses need to be prepared for the individuals involved in this review and planning in order that they can understand the concepts of corrective action and problem analysis.

## Implementation

The history of problems and their causes needs to be kept alive. Design engineers or data information systems people, for instance, need to have a catalog of sorts that shows the things that have happened in the past. Many operations make the same mistakes over and over again because the people who have been there a long while have no way of telling the new people about them.

# COMMUNICATIONS

## COMMUNICATIONS A

Information about the progress of quality improvement and achievement actions is continually supplied to all employees.

## Determination

Keeping everybody up to date on the status of the improvement process is something that cannot be left to chance. Communication within the culture system of the company has to be continually fed in a way that lets people know that many things are happening.

## Education

It is up to the quality improvement teams to solicit information about improvements achieved and to keep it coming into the communications system. This should include events occurring in the work of suppliers and customers. Real-life achievements are the kind that serve as a motivation to other people and the thing that keeps them improving their own comprehension and activities.

## Implementation

In addition to normal company communication material, quality newsletters can be distributed both in-house and at home. One of the best ways of making information available is videotapes, in which the actual improvement can be shown and discussed by the people who took part in it. These have a very high credibility. A video introduced by the president or another senior executive of the company and then carried on by those employees who actually caused the improvement will lead others to make the same sort of commitment.

## COMMUNICATIONS B

Recognition programs applicable to all levels of responsibility are a part of normal operations.

## Determination

Recognition that lasts, that is meaningful, comes as a result of work that is performed to agreed-upon measurements. That is why it is absolutely necessary that the measurement process be set up and established prior to recognition.

## Education

Members of the quality improvement team who are going to determine the recognition needs should have a special course on recognition to make certain that they put things in the proper perspective. Recognition must fit into the culture of the company. The entire company needs an education on what the recognitions are all about, what they mean, why they were established, and by whom they are presented.

## Implementation

The most valuable recognition comes as a result of peer judgment. As much as possible it should be established in that way. It is not necessary to give things of great value in a recognition. One's name in the paper is enough in many cases. However, a full spectrum of appreciation must be available so that a person is not automatically shut out of appreciation once he or she has been awarded something.

## COMMUNICATIONS C

Each person in the company can, with very little effort, identify error, waste, opportunity, or any other concern to top management quickly—and receive an immediate answer.

## Determination

Top management must want to hear from people. If it does not want to hear from people, then it will not. The "My door is always open" type of management is obsolete; it is too passive. Management has to reach out and give people the opportunity to communicate without going through a great many channels.

## Education

A system should be set up so that a person can instigate communication by filling out a form or calling a number or doing something simple, and everyone should know about it. The system should become part of the functioning of the company; it should not be tacked on.

## Implementation

This communication system should be managed by personnel, with an overview from the quality improvement teams to make certain the people are not being hassled. Each communication has to be treated just as if it came from a director of the board.

### COMMUNICATIONS D

Each management status meeting begins with a factual and financial review of quality.

## Determination

Quality is to be first among equals, and it should be first on the agenda. Management must assure itself that the problems are being identified and eliminated, and that the cost of quality is moving lower each time for the right reasons.

## Education

Executives and managers who participate in these status meetings need to understand where the responsibilities for quality and quality improvement lie. In addition to their quality education, they need an understanding of problem identification and elimination.

## Implementation

Charts on the financial status and on problem identification and resolution should be prepared for the status meeting. The same chart should be used every time with a continuing line bringing it up to date. Functions responsible for corrective action should be identified specifically.

## OPERATIONS

### OPERATIONS A

> Suppliers are educated and supported in order to ensure that they will deliver services and products that are dependable and on time.

### Determination

Supplier quality management sessions are held so that each supplier has the opportunity to participate. These sessions are all carefully prepared for the purpose of establishing a positive and permanent communication between suppliers and the purchaser. Those companies desiring to become suppliers also should come to these meetings.

### Education

Supplier quality management workshop needs to be conducted for all personnel who are going to be involved in relationships with suppliers. This is to guarantee that they have a common language and that they are treating the suppliers in the proper way. It is necessary for all of them to recognize that half the problems are caused by the purchaser and that the supplier should be looked at as an extension of our own operation.

### Implementation

After a workshop education, supplier quality management sessions should be held on a regular basis beginning with the suppliers who produce the material of most significance or value.

### OPERATIONS B

> Procedures, products, and systems are qualified and proven prior to implementation and then continually examined and officially modified when an opportunity for improvement is seen.

## Determination

The overall company practice has to be that we do not use any procedures, products, or systems that have not been proven. This practice has to be followed explicitly, because if it is not people feel encouraged to take "calculated risks." There are very few product safety problems that were not caused by a calculated risk.

## Education

Emphasis on procedural and product qualification should be an integral part of the employee education system discussion courses. One purpose of this is to help people recognize that it is important to take procedures seriously as they are developed. When procedures are casual things that are only used to settle arguments, a system is headed for big trouble.

## Implementation

Qualification of products and improvement of procedures should be done to the satisfaction of the quality function. It does not have to be done *by* quality. However, the basic rule is that the qualification for improvement has to show that the procedure or product does indeed fulfill the basic requirements that were intended from the start. If it does not, then they have to back up and redo it or come up with a set of requirements that is acceptable to everybody that the product or procedure meets.

### OPERATIONS C

Training is a routine activity for all tasks and is particularly integrated into new processes or procedures.

## Determination

It is not enough to just write a procedure on how to do something or install a process or even come up with a new product. Those who have to use it or make it or do something to it have to learn their role. One portion of every development activity is training that permits it to be implemented all through the system. It extends from a new form that salespeople have to fill out to the electronic connection of the assembler or the new software insert of the computer operator.

## Education

Training must be done in a formal way so people will know that it is going on. This means that classes must be structured and held. Courses must be documented in some way, and some record must be kept of those who qualify. Once training has been implemented, it is necessary to follow up and see whether the students are indeed practicing what they learned. Those who are having difficulty need to be sent to "reform school."

## Implementation

The funds necessary for training and all educational activities as they apply to a company's work should be included as a normal part of operations. Some education can be conducted inside; some must be conducted outside. The question is one of expertise and convenience. However, the quality improvement team must conduct a continuing overview to make certain that training as a preventive activity is conducted normally.

## POLICY

### POLICY A

The policies on quality are clear and unambiguous.

## Determination

The purpose of the quality policy is to make certain that everybody understands it. All the policies should avoid weasel-words or words that can be used to permit nonconformance to exist at all. Many examples of basic policies are included in other sections of the book. However, there need to be basic policies on some of the operational activities, such as qualification testing, product safety, field service reporting, and so forth. All policies must pass the test of clarity.

## Education

The policies should be taught as part of the overall employee education system. Then the quality improvement team should ensure through continual checking that they are understood as well as utilized. The policies need to be made visible so people can read them, hear them, and know they really exist.

## Implementation

Policies should not be hidden in books for just a few people to read. They should be posted on the wall. They should be painted on the water tower. They should be wherever it is necessary for them to be in order for everybody to clearly understand they are the way the company operates.

### POLICY B

The quality function reports on the same level as those functions that are being measured and has complete freedom of activity.

## Determination

Management has to have made up its mind that it does not intend to be surprised again about the deterioration of product or service quality. It always has to have an objective, continuous, systemwide measurement, a reporting system that lets it know where things stand. This quality function must be staffed by people dedicated to that goal who are management oriented, who have some professional training, and who are working on the basis that their responsibility is the integrity system of the company. This function should report at the same level as the functions that are being measured and have freedom of expression and of participation in the management system. It does not have to be a large function. Just a few well-oriented people working on prevention is all it takes.

## Education

All the members of the quality function must have attended executive and management education and, in addition, they should be exposed to the counseling of people who have many years of experience in causing prevention. They must be counseled not to become involved in troubleshooting, problem solving, or any of those things that keep them away from preventing future problems.

## Implementation

The quality function must work with every other single function in the company to help install a measurement system and reporting entity so that they never get surprises in terms of conformance. This places the quality function in primarily an educational posture, in that it is helping others to do what they want to do anyway.

## POLICY C

Advertising and all external communications must be completely in compliance with the requirements that the products and services must meet.

## Determination

The policy has to be very clear that the company never promises something it cannot do or advertises something it cannot produce. Many companies do this unknowingly, because they feel that much of advertising or communication is exaggeration anyway. However, employees watch these ads, as well as customers and potential customers. If the employees see that a company is promising something it cannot produce, then they will lose faith in management's commitment to quality.

## Education

Representatives of the advertising agency and the advertising function need to attend executive education and to understand all that the policies imply. They can then construct an advertising ethic that everybody can agree to that will prevent problems from happening in this area.

## Implementation

The quality improvement team should review some of the past advertising and external communications of the company and make positive comments if any of those violated what is now the policy of the company. This learning process will help the advertising and external communications people to put together their ethic.

# 18
# How Come Very Little Ever Improves All by Itself?

I make a hundred or so speeches a year. Many are to Quality College attendees, many to client companies, and many to large groups such as industry associations. I really enjoy being a banquet speaker because the room is usually full of happy people of both sexes. It gives me the challenge of seeing whether their attention can be captured and held.

Since each audience is different, each speech is different. The basic message may be the same, but each presentation is unique. However, in every case I like to ask why it is so difficult to get improvement even though everyone is all for it. I have never met anyone who was against quality or for hassle. Companies spend incredible amounts of money and effort on improvement, yet they gain little from it. I see companies with well-meaning staffs struggling to cause improvement and getting very small results. (Most of the reasons for this have been covered in earlier chapters.) Much of what they do produces little change. They buy tapes or books on the subject, they run classes—adjusting philosophies that took a lifetime to develop so they will fit that company's culture—they create motivation and communication programs, and, in general, they apply all the known principles of improving a function. They work very hard, and they are dedicated to the core. It is a terrific waste of people and talent.

American business has had all the problems it has with quality because it doesn't take the subject seriously. We have to be as concerned about quality as we are about profit. That may not come about in this generation.

The business media never write a serious article about quality as a management subject. They only write about techniques that can be applied to low-level jobs and cause improvement. They have not bothered to understand quality as "conformance to requirements." They do not recognize that the primary job of management is to establish requirements, provide the wherewithal to meet them, and then spend all its time getting the requirements met. Because they treat quality as something abstract, like an American League pennant race, management sees it the same way. American management learns its art from the media and the business schools.

A few enlightened business school professors have begun to realize that the management of quality is more than applying techniques. However, I have not seen them address the subject of hassle yet. Several have asked for material we use to teach our clients. They adjust it just enough to reduce its effectiveness, but it is a beginning.

These thought leaders head a general business management culture that says quality improvement and hassle elimination are special subjects outside the day-to-day struggle of running the company. Thus the professionals of human resources and quality control are left with the task. You don't have to be around these folks very long before you recognize that they consider a little hassle and a little nonconformance as part of the human equation. "It is just some inevitable thing that we will try to minimize as much as possible."

Profit is another thing altogether. Profit is seen as the prime responsibility of the top management. It is one of the first things board members talk about when they come together formally or informally. It is the prime measurement of the effectiveness of the CEO. Everyone is absolutely determined that profit will reflect honor on him or her. A lot of things—all proper—are done to make certain this happens.

Quality will never cease to be a major problem until management believes that there is absolutely no reason that we should ever deliver a nonconforming product or service to our customers. When management respects the rights of the customers exactly the way it respects the rights of the banks and stockholders, then quality will happen all the time.

When honoring the right of employees to be free of hassle is considered as important as increasing sales, then hassle will be eliminated.

There are certain things that can be handed off and work very well, but none of them are philosophical items. The basic purpose of the organization is revealed by what the management considers vital. I can think of only one or two companies in the U.S. where quality is something management would "die" for.

Fighting the established thinking has always been a real difficulty, since very few will listen. In the next decade more products, sectors, and companies will die from lack of quality than from lack of money. There will be a complete awakening to the significance of doing what we said we were going to do for our customers. Those who wake first will grab the market. Those who slumber too long will work in other people's branch offices.

With any luck I will have gone to my reward by then, but I will not be thinking, "I told you so." It has been known for generations that organizations that do what they promise to do, and take good care of their employees in the process, always come out on top. When that doesn't happen, it is because either the management is too self-oriented or it just does not understand. The preceding chapters have provided enough information for the beginning of understanding at least.

If understanding is available, and the need is apparent, how come improvement is not happening? The answer is that it is not in the best interests of those who run the place to cause improvement at this time.

That is something they will have to explain to you. I hear reasons all the time, and some of them are interesting. Very few are believable. But they are real. If this is the case in a company, no improvement effort will succeed.

The producing of defect-free products and services on time is mostly caused by the minds of those who hold the strings of power. If something you want to happen is not happening, follow the string back to its origin. It might terminate in the office of someone you know. Perhaps even yourself.

When you know who needs to be corrected, you can cause the correction. Improvement is as much fun as it is reward.

# Guidelines for Browsers

Controls help, but it is very difficult to make motivated management people read the procedures, let alone follow them. Having a large book of policy and practices never saved any company from disaster.    7

Antibodies must be built into the management style that operates the company.    7

All nonconformances are caused.    7

Anything that is caused can be prevented.    7

Determination evolves when the members of a management decide that they have had enough and are not going to take it anymore.    7

Education is the process of helping all employees have a common language of quality, understand their individual roles in the quality improvement process, and have the special knowledge available to handle antibody creation.    7, 10

Implementation is guiding the flow of improvement along the "yellow brick road."    10

Most companies concentrate on implementation before they do anything about the other two.    10

When problems arise, it is entirely normal to reach out for the current fad.    10

To dehassle forever, it is necessary to change the company's culture, to eliminate the causes that produce nonconforming products and services.    10

A short-range solution had been applied, and short-range results had been achieved.    11

If you think you don't eat much, write it all down for fourteen days.    11–12

Getting weighed ten times a day in different ways would not reduce the weight a bit.    13

Getting employees "turned on" has become a major industry.    14

"Why do we need a special program to motivate our people? Didn't we hire motivated employees?"     14

There is nothing wrong with these programs except that they are all aimed at the bottom of the organization.     15

Employees are turned off to the company through the normal operating practices of the organization. The thoughtless, irritating, unconcerned way they are dealt with is what does it. They feel they are pawns in the hands of uncaring functional operations.     15

The performance review, no matter how well the format is designed, is a one-way street.     16

Dishonest evaluations show people that the company has no integrity, doesn't trust a system it forces on them, and doesn't really care about talent if it exists.     16

Expense accounts are another prime opportunity to convince employees that their concern is of no concern.     16

Accounting gets to determine whether the traveler has complied or not.     17

The difficulty of the form is a key to the success of the mission. The mission is to make traveling difficult.     17

"Staff meetings are the same everywhere. The boss talks about what he wants to talk about as long as he wants to talk about it, and then the meeting is over."     18

There is absolutely nothing more demotivating or demeaning to a budding executive than to have to go to meetings where the assigned role is to be a faithful listener.     18

Eighty percent of the talking is done by 20 percent of the people.     18

Meeting culture drives more talented people out of companies than any other hassle, in my opinion.     18

Being an employee in a hassling company is a lot like living at home after you grow up and having your parents decide all kinds of things for you.     19

What we do to ourselves is our business, and we probably get what we deserve. However, there is no good reason why others should do unto us.     19

A "hassle" company is one in which management and employees are not on the same side. The "hassle-free" company is one in which all employees are together and there are no sides.     19

"Hassle-free" offers pleasant working relationships, a smooth system, and happy employees.     19

"Hassle" means that the people inside the company spend more time working on each other than they do making something happen.     19

Of course, there are pathological situations that are not very fixable: the boss who is so concerned with being important that no human communication is possible; the devious or corrupt leader; the power struggle that becomes open warfare in which villages and villagers are destroyed.     19–20

The actions that hassle employees and create a negative atmosphere are usually not big items.     20

In a hassle-free company the employees have confidence that the management respects them and needs their output. They know that the requirements of the job are clearly stated, and they have had the opportunity to make inputs to that statement.     24

They recognize that management is committed to performing to those requirements and takes them seriously. They see that recognition is accorded to those who do well and help is given to those employees having difficulty.     24

Making all this come about requires that the company be involved in a lifelong process of quality improvement.     24

The value of the hassle-free company is obvious for the most part. One aspect that may not leap immediately to mind is that it is a great place to look like a sensational manager.     25

Hassle means different things to different people. But to me it is the unnecessary difficulties or harassment placed in the path of someone trying to do a reasonable thing.     34

"Stopped the line? Stopped the line?" roared Spellman. "Are
there no rework areas? What kind of place are we running
around here? If we stop every time some little thing goes wrong,
we won't have any product to sell. That is what we have field
people for."   37

"Tell them to hire trained people. We're not a university."   37

"You people are going to have to quit trying to gold-plate
everything around here. We can't afford to spend money on
quality."   37–38

". . . the way to make money is to give the customers what
they were promised and eliminate the hassle for the
employees."   41

"Impractical? But we always concentrated on the practical. I
thought the idea was to cause the most efficiency. And certainly
quality is part of that. We spend a lot of time deciding how good
things have to be."   41

"We'd have to do ten times as much checking to see that no
defects happen. It would be impossible."   43

"Let's get all the workers together, tell them that from now on
we are all going to do everything right the first time, invite the
customer to speak to them, have a band—and let them think we
are going to do something different."   43

"Why do you think it would be so impossible to prevent
problems?"   43

"Causing quality to happen is beyond the reach of the quality
control people unless management completely understands their
role."   44

". . . senior management is 100 percent responsible for the
problems with quality—and their continuance."   44

"One primary measurement of quality is the price of
nonconformance. That includes all the expense of doing things
wrong."   44

". . . the price of conformance—what is spent on inspection,
testing, and education."   45

"Unquality costs money. Fixing, correcting, scrambling—all those things cost money." 45

"I don't understand all this fuss about quality. It has always been around; it has always been handled one way or another." 45

"So the 60 people in the quality circle's program are certainly a valuable asset. But that hardly represents any kind of commitment to improvement." 46

". . . you are going to have to stop deluding yourself about quality. CC has a growing reputation for shoddy service and merchandise." 46

"My biggest problem is suppliers that give us services, and products also, that just are not right. I spend most of my time with the lawyers and contract people trying to get all this fairly adjusted. And it is getting worse." 47

"Meeting the schedule is my biggest problem. We are having to make five of everything to deliver four." 47

"Our inspection effort is spread very thin since we have to do a lot more checking now." 47

"My biggest problem is keeping our products up and operating in the field." 47

"All my time is spent traveling from one store to another urging the managers to work harder on bringing customers into the stores. They keep on me to get them better merchandise and more adept clerks." 48

Overall the consensus was clear that this was just a result of the social and business level of the day. They were probably doing everything that could be done. 48

"Quality improvement is a process, not a program, and it takes a long time for it to become a normal part of the scene." 50

"The first thing that has to happen is for all the 'thought leaders' of the company to understand quality the same way." 50

"The level of people I'm thinking about need outside education. They need to understand the Absolutes of Quality Management. They have to get in touch with reality, and so do you." 50

"Convention says that quality is goodness and therefore something vague; reality says that quality is conformance to requirements and therefore very specific." 50

"Convention says that quality is achieved through inspection and testing and checking; reality says that prevention is the only system that can be utilized." 50

"Convention says that the performance standard for employees should be acceptable quality levels or 'That's close enough'; reality says it must be specific, like error-free, or having zero defects." 50

"Convention says that quality should be measured by indexes and comparisons; reality says that we should calculate the price of nonconformance." 50

"I think error-free probably is a little strong. That sounds unrealistic." 50

"The comprehension to bring about this improvement has been known for some time. The willingness of management to do it has not. They are a stubborn breed." 51

"The quality policy of this company is that we will deliver defect-free products and services to our customers, both internal and external, on time." 51

"Quality is not just making things like requirements and being steadfast in insisting on doing things right. It isn't a functional thing at all. Actually it has to do with the way the company is run overall. It takes the combined actions of every person in the company to cause the company to operate properly." 51–52

"What we are going to make happen around here is a hassle-free company. We are going to learn how to do all the things we do in an orderly, energetic, and accurate fashion." 52

Companies don't do well with quality because they are just not determined enough. 53

The companies that don't get much improvement, even though they appear to be determined, have common characteristics: 53

1. The effort is called a program rather than a process.    53

2. All effort is aimed at the lower level of the organization.    53

3. The quality control people are cynical.    54

4. Training material is created by the training function.    54

5. Management is impatient for results.    54

These characteristics, and a few more, show up in the poorly run quality improvement process. They occur because the entire event has not been thought out and taken seriously enough.    54

I never have felt that you could "motivate" anyone for more than a few days.    55

The company wasn't trying to do wrong; it just felt some of the requirements were not that critical.    55

Here was a case where the exact same organization produced two opposite results. The only difference, outside the two fired hardheads, was the leadership.    56

The GM made it crystal clear, day after day, that he was determined to produce quality in the truest sense of the word.    56

American management has almost worn a rut between here and Japan searching for the secret of Japanese quality.    56

They come back with reinforcement for the idea that the problem is the worker and what is needed is some way to get American workers to act like Japanese workers.    56

But little improvement will happen until the real problem is dealt with. The real problem is that management doesn't take the product and service requirements seriously.    56

If you look behind the Japanese success, or the success of the many American companies that have quality products, you will find one secret: they take the requirements seriously.    56

They create them with care; they meet them with care.    56

When management insists on conformance to requirements and

provides the participation necessary for prevention to happen, it becomes a different world.     57

It is not enough to look determined and act determined. The subject we are being determined about has to be clear in the minds of all involved.     57

The credibility of the commitment is the biggest single problem for management; it has to be reinforced all the time.     57

The actions and the lifestyle have to be visible.     58

It is difficult to reach the mind of someone who is enthusiastically agreeing with you.     58

The First Absolute: The Definition of Quality Is Conformance to Requirements     59

Quality improvement is built on getting everyone to do it right the first time (DIRFT).     59

Management really has three basic tasks to perform: (1) establish the requirements that employees are to meet, (2) supply the wherewithal that the employees need in order to meet those requirements, and (3) spend all its time encouraging and helping the employees to meet those requirements.     59

Hassle comes about because of vacillation in management's dedication to the policies and processes.     59

We had agreed on the requirements and they were ours.     60

Such activities cost service companies a conservative 40 percent of their operating costs.     60

The cause is management's definition of what quality is: goodness. Nobody knows what that means except the speaker.     60

Quality has to be defined as *conformance to requirements*.     60

The determined executive has to have a brain transplant where quality is concerned.     60

It is not necessary to shut the company down to prove you are a determined executive.     61

If we do not take the requirements seriously, then we will not perform the task well.   63

Taking requirements seriously is the first act in improvement.   64

The first Absolute of Quality Management is:

QUALITY HAS TO BE DEFINED AS CONFORMANCE TO REQUIREMENTS, NOT AS GOODNESS.   64

Managers tend to get very worried when the subject of setting requirements comes up. They immediately visualize thousands of little "Do this and do thats."   64

Requirements, like measurements, are communications.   65

The Second Absolute: The System of Quality Is Prevention   66

The most visible of the expenses of conventional quality practice lie in the area of appraisal.   66

Appraisal, whether it is called checking, inspection, testing, or some other name, is always done after the fact.   66

Appraisal is an expensive and unreliable way of getting quality.   67

Prevention is something we know how to do if we understand our process.   67

If a salesperson is driving from a strange airport into a strange town, it is best to ask for directions prior to heading out on the highway.   67

If a painter wants to match a color, it is best to take a sample to the paint mixing store.   67

If a restaurant owner wants fresh eggs every morning, it is necessary to locate someone who sells fresh eggs and have them delivered on time.   67

It is hard not to agree with these common-sense actions. It is much too late to check sizes when the boxes arrive, to open each egg and look inside, to run back and forth trying to remember colors, or to peer at a road map as an eighteen-wheeler rides the rear bumper.   67

Prevention is one of those things that business people just don't talk about.   68

The secret of prevention is to look at the process and identify opportunities for error.   68

The development and preparation of that information is a major task in itself.   68–69

Massive opportunities for error were eliminated, and that is what prevention is all about.   69

SQC is made out to be very complicated and difficult to do, but there really isn't that much to it.   70

Executives do not realize the effect their personal actions have on the processes of their company.   70

The best laid plans for prevention can be undone by a careless executive policy.   70

"First, remember that all the problems are caused by management. Second, run the line according to the requirements that have already been established. If something goes out, stop and fix it. It will all run smoothly soon."   73

The second Absolute of Quality Management is:

THE SYSTEM FOR CAUSING QUALITY IS PREVENTION, NOT APPRAISAL.   73

The Third Absolute: Performance Standard Is Zero Defects   74

A company is an organism with millions of little seemingly insignificant actions that make it all happen.   74

The performance standard is the device for making the company happen by helping individuals to recognize the importance of each one of these millions of actions.   74

Companies try all kinds of ways to help their people not meet the requirements.   75

Payroll doesn't make mistakes.    75

The reason payroll does so well is that people just won't put up with errors there. They take it very personally when something is wrong with their paycheck.    76

Conventional wisdom says that error is inevitable. As long as the performance standard requires it, then this self-fulfilling prophecy will come true.    76

Only someone who has had the job of causing quality in an organization can realize the importance of a specific performance standard.    76

Unfortunately, Zero Defects was picked up by industry as "motivation" program.    77

Companies have elaborate reporting systems to show that they are improving. They have advertising programs that show their people working hard on quality. The only thing they don't have is error-free products.    77

"Somewhere in the world," he said, "there is a quality manager who can get me products and services with no problems in them. I sure would like it to be you."    81

People do make mistakes, particularly those who expect to make some each day and do not become upset when they happen. You might say they have accepted a standard that requires a few mistakes in order to be certified as a human.    82

. . . then errors must be a function of the importance that a person places on specific things.    83

Mistakes are caused by two factors: lack of knowledge and lack of attention.    83

People will perform to the standard they are given, provided they understand it.    84

When the standard is specific like Zero Defects, defect-free, or DIRFT (for Do It Right the First Time), people will learn to prevent problems.    84

All the results in a company are made by people.    84

The third Absolute of Quality Management is:

THE PERFORMANCE STANDARD MUST BE ZERO DEFECTS, NOT "THAT'S CLOSE ENOUGH."    84

The Fourth Absolute: The Measurement of Quality Is the Price of Nonconformance    85

The main problem of quality as a management concern is that it is not taught in management's schools.    85

The reason is that quality is never looked at in financial terms the way everything else is.    85

Cost of quality is divided into two areas—the price of nonconformance (PONC) and the price of conformance (POC).    85

The price of nonconformance can be used in two ways, one, as a whole to track whether the company is improving or not, and two, as a basis for finding out where the most lucrative corrective-action opportunities reside.    86

Collecting the cost of quality is not a difficult task, but it very rarely gets done in a company.    86

Take everything that would not have to be done if everything were done right the first time and count that as the price of nonconformance.    86

The fourth Absolute of Quality Management is:

THE MEASUREMENT OF QUALITY IS THE PRICE OF NONCONFORMANCE, NOT INDEXES.    86

I've always believed that the standard process of business education does not do a good job of transferring understanding to the student.    87

Techniques developed by educators to guarantee that the student will learn a specific bit of information pile up on the rocks of human individuality.    87

There is just no standard way to guarantee comprehension.    87

It is not enough to make a cookbook for something like hassle elimination, because that will just add another level of hassle to the operation. People will set up procedures to implement the items in the cookbook, and pretty soon compliance with the procedures, or lack of compliance, will become a bigger hassle than it all was in the first place.     87

Producing a hassle-free company requires the continual transfer of information from person to person.     87

Everyone talks about the need to do things right the first time and no one really wants to do things right the second time. Yet in real life it may be the third time before anything gets done properly.     87–88

A solid understanding of a subject means comprehension.     88

Quality, like golf, is one of those areas where one can get along for years with just a veneer of information.     88

The Absolutes of Quality Management must be understood by every single individual. These are the common language of quality.     88

The fourteen-step process of quality improvement needs to be understood by the management team since they are responsible for making it happen.     88

The individual's role in causing quality must be understood by each and every person in the company.     88

The overall educational aspect requires executive education, wherein senior management can learn its role; management education, wherein those who must implement the process learn how to do it; an employee education system, wherein all the employees of the company learn how to comprehend their roles; and workshops, wherein special functions such as purchasing, accounting, quality, marketing, and so forth can learn how to do the individual and special things that are important in their world.     88

The purpose of executive education is to help senior people understand their role in causing problems and then causing improvement in the quality process.     89

In management education, all the content from executive education is covered with the addition of several items.    89–90

It is important that this group recognize they have to present the quality improvement case on a continuous basis.    90

Employee Education. The other 98 percent of the people in the company receive their primary quality education from student notebooks.    90

We've learned to put together each segment in a standard way. One, material to read prior to coming to class on the subject. Two, a video, usually fifteen minutes, that explains concepts to be discussed during that module. The video is constructed using actors and original scripts. Three, a workshop in which the concept can be applied to something with which the student is familiar. Four, discussion in which the instructor leads the students in determining how the concept applies inside that particular company. Fifth, a work assignment.    90

The entire education process can be summarized in what I call the "six C's": Comprehension, Commitment, Competence, Communication, Correction, and Continuance.

*Comprehension* is the understanding of what is necessary and the abandonment of the "conventional-wisdom" way of thinking.    92

*Commitment* is the expression of dedication on the part of management first and everyone else soon after.    92

*Competence* is the implementation of the improvement process in a methodical way.    93

*Correction* is the elimination of the opportunities of error by identifying current problems and tracking them back to their basic cause.    93

*Communication* is the complete understanding and support of all people in the corporate society including suppliers and customers.    93

*Continuance* is the unyielding remembrance of how things used to be and how they are going to be.    93

All the employees who need to know are taught until all of them understand. Then the ones who didn't need to know then, but do need to know now, are taught.  98

The process of installing quality improvement is a journey that never ends.  98

Nothing happens just because it is the best thing to do, or just because it is worthwhile.  98

Changing a culture is not a matter of teaching people a bunch of new techniques, or replacing their behavior patterns with new ones. It is a matter of exchanging values and providing role models. This is done by changing attitudes.  98

All those terrible things that people are doing, the hassles that cause defects, are done with the best of intentions.  99

The culture we have now was caused.  99

Although I have been working with these blocks for twenty years I still learn about them every day.  99

In talking with companies who are experiencing improvement the executive has realized that senior management is the key to the solution, as well as the cause of the problem.  100

Only when the management team becomes educated and sets out on its mission of changing the culture of the company can it hope to reach the rewards such a change produces.  100

The problem used to be: "How do we get management interested enough in quality to do something about it?"  100

Now the problem is: "How do we get the people to believe that we are really going to do something and stick with it?"  100

The culture of the company is going to change only when all employees absorb the common language of quality and begin to understand their individual roles in making quality improvement happen.  100

The way senior management acts toward accepting anything less than the requirements lays the foundation of this faith.  101

First, a corporate policy on quality needs to be issued. This policy should make it clear that the commitment is real and understandable. It must have no weasel words in it.   101

Second, quality should be made the first item on the agenda of the regular management status meetings. It should come before finance and be discussed in specific terms.   101

Third, the CEO and the COO need to compose clear quality speeches in their mind and, as they go around the company, deliver them to everyone whose path they cross.   101

"We will deliver defect-free products and services to our clients, on time."   101

The message has to be clear from top to bottom. "We will take the requirements very seriously. If we don't need something then let's officially change the requirement. But please don't ask me to agree to deviations. We need to spend our time learning how to make things right."   104

Management commitment is tested and tested until it can be assumed.   105

The quality improvement team requires a clear direction and leadership. Otherwise people can get so involved with strategy, and the selection of the team, that they forget what the team is for.   106

The purpose of a team is to guide the process and help it along. It is not to clear each action beforehand, to be the all-wise oracle, or to hold things back.   106

The quality improvement team should be made up of individuals who can clear road blocks for those who want to improve.   106

The chairperson of the team should be someone with an easy conversational access to the very top management.   106

Top management, the coordinator, and the team chairperson lay out the overall strategy.   107

The team members all need to have the same educational base

concerning the quality improvement process or it will never get out of the chute. Those who don't understand the concepts will lead the entire effort into a low-level motivation program.     107

The real learning comes from the experiences that the team members themselves have.     107

The best training is playing.     107

Consider Little League baseball. Would the purpose of that effort be just to score runs?     107

The purpose of childhood athletics is to help the individuals learn more about getting along with others, while understanding themselves better.     107

Those who expect quality improvement teams to become corrective-action functions have the same limited approach to management as those who count Little League only by the pennants that are won.     107–108

It's when no one can tell how well you're doing that you get frustrated.     108

It is very difficult to have any kind of conversation about anything that doesn't include several measurements. If this were not so, we would be unable to communicate in definite terms.     108

Measurement is just a habit of seeing how we're going along.     108

Everyone thinks the other people have clear requirements and they alone are left in the dark.     109

The creative person needs to communicate the creation to other people. This can only be done through processes, procedures, and measurements.     109

Many chief executives immediately pounce on the cost of quality as yet another way of measuring the performance of their executives.     110

When a company cost of quality has been identified and fed into the regular management process, it serves as a very good and positive stimulus for the quality improvement process.    110

There's nothing like money to get management's
attention.    110

The most effective quality awareness systems seem to be those
that use existing systems inside a company. Instead of being in a
separate quality newsletter, for instance, quality awareness
becomes part of the regular company newsletter.    111

Awareness is not just making publications and promotions and
so forth; it is spreading information.    111

Most companies feel they have a corrective action system, yet
they still have a lot of problems that don't seem to get solved in
any reasonable length of time.    111–112

The real purpose of corrective action is to identify and eliminate
problems forever.    112

Corrective action systems have to be based on data that show
what the problems are and analyses that show the causes of the
problems. Once the root cause has been determined, it can be
eliminated.    113

There are very few causes for error in the supplier-purchaser
relationship.    113

Zero Defects Day is a celebration and will take care of
itself.    114

We developed a complete quality education system that would
provide a standard message and could be taught by anyone who
was trained to use it.    115

People became less tolerant of hassle at all levels of the company
as well, and it began to disappear.    116

There are still those in the quality profession and other isolated
areas who think the purpose of Zero Defects Day is to get all
the employees together so they can sign a commitment to
improve.    116

It is a time to show all the people, face to face, that management
is serious.    116

A great many people very rarely have exciting days at work.    116

Goal setting is something that happens automatically right after measurement.    116

Error-cause removal is asking people to state the problems they have so that something can be done about them.    117

Each of us sights on some other person, consciously or unconsciously, as a reference point.    117

Very few companies recognize their good performers.    118

People don't work for companies; they work for people.    118

I have become even more convinced that money is a very bad form of recognition. It is just not personal enough.    119

The idea of quality councils is to bring the quality professionals together and let them learn from each other.    119

"The only employees not in the service business are those who are professional blood donors; they are a resource."    124

When a customer deals with a hotel he or she meets all the employees at the bottom of the organization.    124

When dealing with a foundry, contact is with top levels of the company.    124

"Field service is profitable only because we don't charge it the full overhead it deserves."    126

"First you are going to establish the environment for causing improvement, by educating people and at the same time formalizing the management commitment, the measurement, and the awareness."    137–138

"Then you are going to examine the whole business of processing the purchase orders. This examination will be successful because the people involved will all understand quality and hassle the same way and will all have an understanding of their personal role in causing quality improvement."    138

"Doing it this way will succeed because everyone will be interested in causing quality improvement instead of fighting it."    138

Every ingredient involved in setting up a permanent system to eliminate hassle and to cause quality improvement requires special attention.    155

The chief executive officer is dedicated to having the customer receive what was promised, believes that the company will prosper only when all employees feel the same way, and is determined that neither customer nor employees will be hassled.    155

The CEO must continually communicate with the customers and the employees of the company to assure them of its determination on this subject.    155

The CEO needs personal awareness to understand the particular role assigned to this job.    156

The CEO needs to see that the corporate policy on quality is issued, is understood, and is communicated to everyone.    156

The chief operating officer believes that management performance is a complete function requiring that quality be "first among equals"—schedule and cost.    156

The COO in many companies represents the voice of practicality.    156

The COO needs the same education as the CEO but must understand the cost of quality and supplier quality management in-depth.    156–157

The COO must make certain to let operating entities know that quality improvement is not optional.    157

The phrase "take requirements so seriously" means that the executives who run marketing, finance, sales, manufacturing, engineering, regional offices, legal, public relations, quality, purchasing, information services, and all the other functions have to recognize that if they are going to have quality improvement in their areas, everyone has to agree on what is going to be done and work to that end.    157

The executive in charge of quality has to tie in all the measurements and reporting with what is supposed to happen in every other function.   157

Each and every function needs to have its own look at the quality improvement process as it applies to that function.   157

Managers need more working-level information about the process than the senior executives.   158

The professional employees know that the accuracy and completeness of their work determines the effectiveness of the entire work force.   159

The creative force of any organization lies in its professionals.   159

One of the things most valuable to the professional is feedback on the service or product from the interim or end customers.   159

The employees as a whole recognize that their individual commitment to the integrity of requirements is what makes the company sound.   159

When all the employees are determined to conform exactly to the requirements and to offer feedback when the requirements are inadequate or impossible to achieve, then hassles will begin to die out and quality improvement will become a fact of life.   160

The entire professional quality function has to become committed to helping cause conformance to requirements, rather than to carefully identifying nonconformances and keeping neat records of them.   160

The quality function must become a well-disciplined and organized operation that carries on the spirit of the quality improvement process without becoming a disciplinary force.   161

Every employee of the company, without exception, must have a complete education in the understanding of quality and what it means to him or her and to the company.   161

Shortcuts always cause problems.    161

The financial method of measuring nonconformance and conformance costs is used to evaluate processes.    162

Evaluating quality according to financial measurements, such as pointed out under the cost of quality, brings it into a very professional and front-level management position.    162

All operating management people certainly must understand how the cost of quality is put together and how it applies to their area.    162

Many companies do not have the slightest idea what happens to their product or their service once it leaves their hands.    162

A simple form that can be understood by everyone and used with little effort must be constructed.    163

The companywide emphasis on defect prevention serves as a base for continual review and planning that utilizes current and past experience to keep the past from repeating itself.    163

Problem elimination needs to be examined as a part of the formal operation of the organization.    163

No problem must be allowed to exist and to recur.    163

The history of problems and their causes needs to be kept alive.    163

Keeping everybody up to date on the status of the improvement process is something that cannot be left to chance.    164

One of the best ways of making information available is videotapes, in which the actual improvement can be shown and discussed by the people who took part in it.    164

Recognition that lasts, that is meaningful, comes as a result of work that is performed to agreed-upon measurements.    165

Recognition must fit into the culture of the company.    165

The most valuable recognition comes as a result of peer judgment.    165

Top management must want to hear from people.    165

Management has to reach out and give people the opportunity to communicate without going through a great many channels.   166

Each communication has to be treated just as if it came from a director of the board.   166

Quality is to be first among equals, and it should be first on the agenda.   166

Suppliers are educated and supported in order to ensure that they will deliver services and products that are dependable and on time.   167

Supplier quality management sessions are held so that each supplier has the opportunity to participate.   167

After a workshop education, supplier quality management sessions should be held on a regular basis beginning with the suppliers who produce the material of most significance or value.   167

The overall company practice has to be that we do not use any procedures, products, or systems that have not been proven.   168

Emphasis on procedural and product qualification should be an integral part of the employee education system discussion courses.   168

Qualification of products and improvement of procedures should be done to the satisfaction of the quality function.   168

Training is a routine activity for all tasks and is particularly integrated into new processes or procedures.   168

It is not enough to just write a procedure on how to do something or install a process or even come up with a new product. Those who have to use it or make it or do something to it have to learn their role.   168

Training must be done in a formal way so people will know that it's going on.   169

The quality improvement team must conduct a continuing overview to make certain that training as a preventive activity is conducted normally.    169

The purpose of the quality policy is to make certain that everybody understands it.    169

Policies should not be hidden in books for just a few people to read.    170

The quality function reports on the same level as those functions that are being measured and has complete freedom of activity.    170

This quality function must be staffed by people dedicated to that goal who are management-oriented, who have some professional training, and who are working on the basis that their responsibility is the integrity system of the company.    170

All the members of the quality function must have attended executive and management education and, in addition, they should be exposed to the counseling of people who have many years of experience in causing prevention.    170

The quality function must work with every other single function in the company to help install a measurement system and reporting entity so that they never get surprises in terms of conformance.    171

Advertising and all external communications must be completely in compliance with the requirements that the products and services must meet.    171

The policy has to be very clear that the company never promises something it cannot do or advertises something it cannot produce.    171

Representatives of the advertising agency and the advertising function need to attend executive education and to understand all that the policies imply.    171

I have never met anyone who was against quality or for hassle.    172

Companies spend incredible amounts of money and effort on improvement, yet they gain little from it.    172

American business has had all the problems it has with quality because it doesn't take the subject seriously.    172

The business media never write a serious article about quality as a management subject.    173

Profit is seen as the prime responsibility of the top management.    173

Quality will never cease to be a major problem until management believes that there is absolutely no reason that we should ever deliver a nonconforming product or service to our customers.    173

When management respects the rights of the customers exactly the way it respects the rights of the banks and stockholders, then quality will happen all the time.    173

It has been known for generations that organizations that do what they promise to do, and take good care of their employees in the process, always come out on top.    174

The producing of defect-free products and services on time is mostly caused by the minds of those who hold the strings of power.    174

# Index

## ABOUT THE AUTHOR

Philip B. Crosby has had 39 years of hands-on experience as a quality management professional and an executive. While with Martin Marietta he created the concept of Zero Defects. During 14 years as Vice President of the ITT Corporation he was responsible for quality worldwide in 87 divisions. In 1979 he founded Philip Crosby Associates, Inc. (PCA), which became the world's largest quality management consulting firm. Each year 20,000 executives and managers attend the Quality College of PCA.

In 1991 he retired from PCA and now serves as Chairman of Career IV, Inc., in Maitland, Florida, spending his time working with those who would become executives, and speaking to groups.

His other books include *Quality Is Free, Running Things, The Eternally Successful Organization, Let's Talk Quality, Leading,* and *The Art of Getting Your Own Sweet Way.*